MW00674087

# NATIONS OF THE WORLD

# EGYPT

*Neil Wilson*

A Harcourt Company

Austin New York
www.steck-vaughn.com

**Steck-Vaughn Company**

First published 2001 by Raintree Steck-Vaughn Publishers,
an imprint of Steck-Vaughn Company.


**Library of Congress Cataloging-in-Publication Data**

Wilson, Neil, 1959-
Egypt / Neil Wilson.
p. cm. - - (Nations of the world)
Includes bibliographical references and index.
ISBN 0-7398-1283-1
1. Egypt- - Juvenile literature. [1. Egypt.] I. Title. II. Nations of the world (Austin, Tex.)

DT49. W565 2001
962 -- dc21

00-062649

Printed and bound in the United States
1 2 3 4 5 6 7 8 9 0 BNG 05 04 03 02 01 00

*Brown Partworks Limited*
Editors: Robert Anderson, Peter Jones
Designer: Joan Curtis
Cartographers: William Le Bihan and ColinWoodman
Picture Researcher: Brenda Clynch
Editorial Assistants: Roland Ellis, Anthony Shaw
Indexer: Indexing Specialists

*Raintree Steck-Vaughn*
Publishing Director: Walter Kossmann
Art Director: Max Brinkmann

*Front cover:* merchant sipping drink, Mousli District, Cairo (right); death mask of Tutankhamun (left); Nile River and towns, near Elephantine
*Title page:* The Sphinx and Pyramid of Khufu, Cairo

The acknowledgments on p. 128 form part of this copyright page.

# Contents

# Foreword

Since ancient times, people have gathered together in communities where they could share and trade resources and strive to build a safe and happy environment. Gradually, as populations grew and societies became more complex, communities expanded to become nations—groups of people who felt sufficiently bound by a common heritage to work together for a shared future.

Land has usually played an important role in defining a nation. People have a natural affection for the landscape in which they grew up. They are proud of its natural beauties—the mountains, rivers, and forests—and of the towns and cities that flourish there. People are proud, too, of their nation's history—the shared struggles and achievements that have shaped the way they live today.

Religion, culture, race, and lifestyle, too, have sometimes played a role in fostering a nation's identity. Often, though, a nation includes people of different races, beliefs, and customs. Many have come from distant countries, and some want to preserve their traditional lifestyles. Nations have rarely been fixed, unchanging things, either territorially or racially. Throughout history, borders have altered, often under the pressure of war, and people have migrated across the globe in search of a new life or of new land or because they are fleeing from oppression or disaster. The world's nations are still changing today: Some nations are breaking up and new nations are forming.

Egypt is the oldest nation state in the world. Its monuments have been famed for millennia as wonders of the ancient world. The country today shares many qualities with ancient Egypt, not least its reliance on the Nile River and the communities based around it. As in the past, Egypt forms a natural bridge between Africa and the Middle East. But today Egypt is also important politically as the first Arab country to make peace with Israel and in its role as diplomat between the Arab states and the West. This role has brought the country much-needed aid as it attempts to solve its problems of overcrowding and poverty.

# Introduction

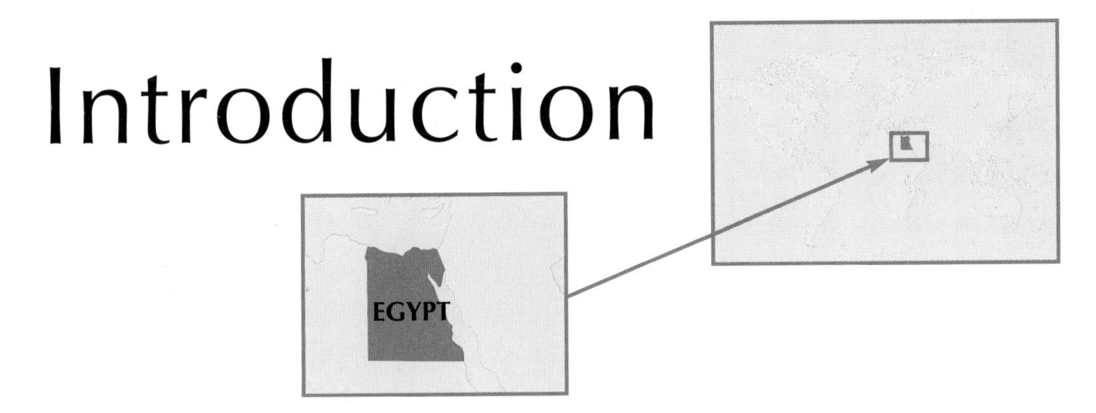


The ancient Greek historian Herodotus once called Egypt "the gift of the Nile." There are few countries in the world that owe their existence to a single geographical feature, but without the waters of the world's longest river, Egypt would not exist.

Seen from an airplane flying between Egypt's capital, Cairo, and the country's ancient capital at Luxor, the miracle of Egypt is plain to see. The Nile and its floodplain support a narrow ribbon of green, rolled out across a vast expanse of barren, sand-swept desert. This fertile strip of land has supported human habitation for at least 7,000 years and nurtured one of the oldest and most magnificent civilizations the world has ever known.

Today, Egypt is the most heavily populated nation in the Arabic-speaking world and is a major political power in North Africa and the Middle East. As one of the first nations to embrace Islam and the home of the oldest Muslim university in the world, it is also an important center of the Islamic religion.

The official name of the country is the Arabic Republic of Egypt (*Jumhuriyat Misr al-Arabiyah* in Arabic). The name Egypt comes from the ancient Greek name for the country, *Aigyptos*. This in turn was derived from the ancient Egyptians' name for their capital city of Memphis, *hut-ka-ptah*, which meant "Temple of Ptah" (Ptah was an Egyptian god).

***Egypt's main geographical feature, the Nile River, supports lush vegetation, but away from the river the Egyptian landscape is mostly arid desert.***

FACT FILE

- Cairo, the capital of Egypt, is the largest city in Africa, with a population of around 9.7 million. It is the most important city in the Arab world and a major business center.
- Egypt is the world's oldest nation-state and has the longest recorded history in the world. This began with the unification of the kingdoms of Upper and Lower Egypt under King Menes around 3100 B.C.
- Egypt is the most populous Arab country in the world and after Nigeria the second-most populous country on the African continent.

*The Egyptian pound is divided into 100 piastres. One piastre is divided into ten millim.*

Egypt's ancient monuments, which include vast pyramids, temples, and tombs, are the cultural legacy of the 3,000-year-long reign of ancient Egypt's rulers, called the pharaohs. They are among the most impressive and fascinating in the world and have inspired so many scholars that "Egyptology" is recognized as a distinct field of study within archaeology. These monuments are also one of the world's leading tourist attractions.

Egypt is a meeting place, too, of West and East, ancient and modern. In the sprawling capital, Cairo, white-domed mosques stand beside modern office buildings. The sound of a muezzin (a Muslim crier) calling Muslims to prayer at specific times of day echoes through streets noisy with automobile horns and pop music. Outside the city, mud-brick villages stand in the shadow of looming pyramids and ruined palaces, crowded with tourists from all over the world. Only deep in the desert, where the wandering bedouin live, does Egypt seem untouched by time.

Egypt's location is seen by world powers as being of great strategic importance. This has allowed the country to obtain military and economic aid from the West, which replaced earlier aid from the former Soviet Union. The country's relations with its Arab neighbors were strained when it became the first Arab country to sign a peace agreement with Israel. Egypt has since worked hard to improve ties with the Arab countries and was readmitted to the Arab League in 1989. In 1990, Egypt led Arab opposition to the Iraqi invasion of Kuwait.

*The Egyptian flag has three bands of red, white, and black. In the center of the white band is the national emblem—a shield superimposed on a golden eagle above a scroll bearing the name of the country in Arabic.*

## POPULATION DENSITY

Egypt's population distribution is one of sharp contrasts. The vast majority of the country's population is clustered around the Nile Valley and the Nile Delta. Within this narrow strip of land the density increases further around the center at Luxor and around the capital, Cairo. Outside of the Nile Valley, in the desert regions, the population density falls to among the lowest in the world

PERSONS

| Per sq. mi | Per sq. km |
|---|---|
| 500 | 200 |
| 1,500 | 600 |
| 2,500 | 1000 |

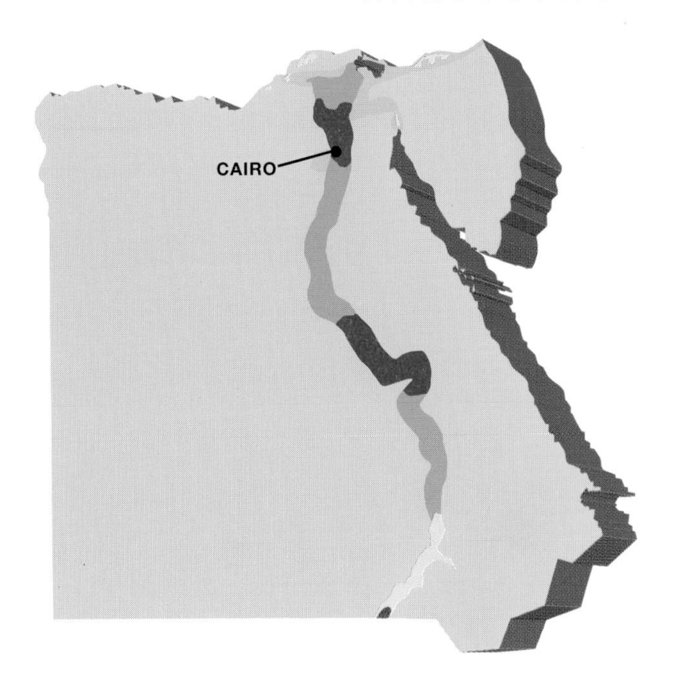


## PEOPLE

Although Egypt has a population of around 66 million, most of the country is uninhabited or only very sparsely populated. Some 98 percent of Egyptians live in the cultivated areas of the Nile Valley and Delta, which account for a mere 6 percent of the country's land. The population density in parts of the Nile Valley is as high as 4,000 people per square mile (1,558 per sq. km)—among the highest in the world. By comparison, New York, Los Angeles, and Chicago have population densities that barely top 3,000 people per square mile (1,150 per sq. km). On the other hand, the desert regions on either side of the Nile Valley are largely uninhabited and extremes of temperature and climate make them hostile to human settlement. The small population of nomads here is clustered around a number of oases, fertile areas of vegetation fed by underground sources of water.

***While the rate of increase has slowed, Egypt's population is increasing by about a million people per year.***

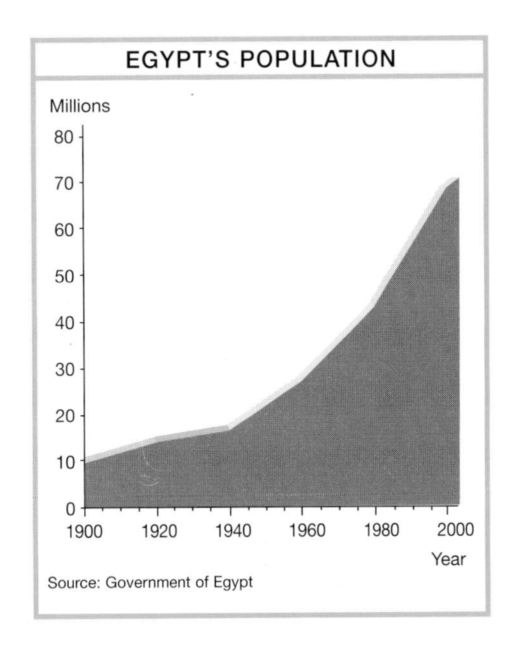

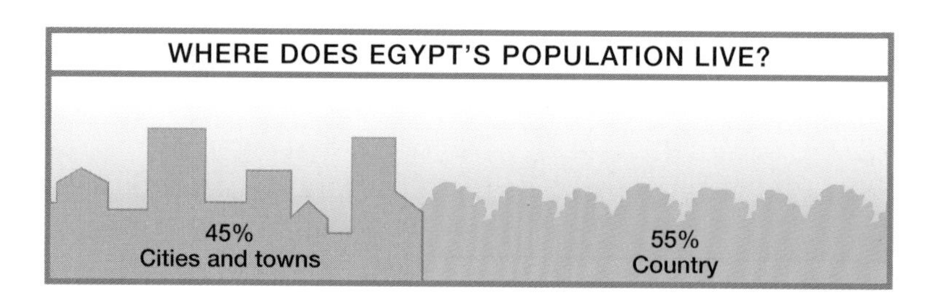


**Right: The Egyptian population is almost equally divided between town and city. Below: The population is very young and overwhelmingly of one race.**

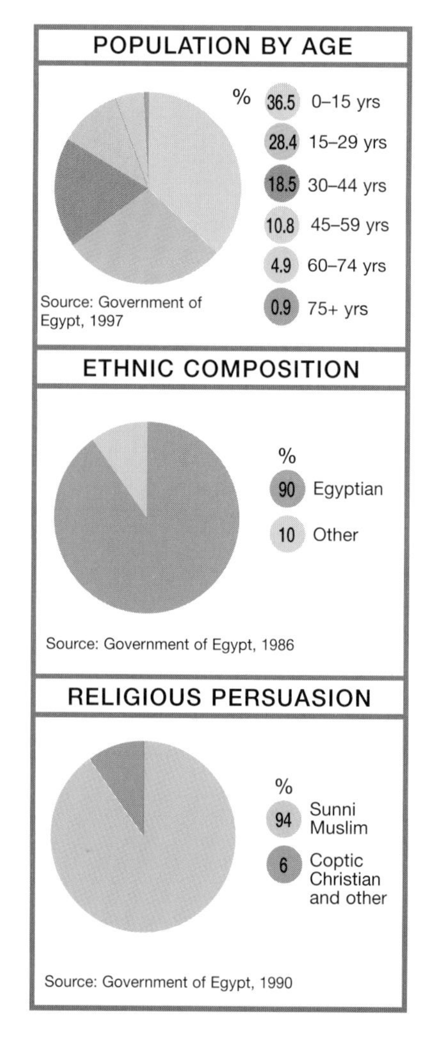


Most Egyptians are Eastern Hamitic, meaning they belong to an Asiatic-African language group from northern Africa. The group consists of Egyptians, bedouin (a nomadic or wandering people, *see* p.115), and Berbers (a group of peoples living in northern Africa from Morocco to Egypt). Most people in Egypt are descended from the native people, the ancient Egyptians, or from the Arabs, who conquered the area in the seventh century A.D. In Upper Egypt the Nubians are an essentially African people who have intermingled with the Egyptians for thousands of years. In the 20th century they were displaced from their traditional home by the building of the Aswan High Dam (*see* p.84). There are also influences dating from conquests by the Greeks, Romans, and Turks. Some 10 percent of Egyptians are Greek, Nubian, Armenian, Italian, or French.

Egypt's official religion is Islam, which was introduced to the region by the Arabs. Some 94 percent of Egyptians are Sunni Muslims. The bedouin follow a mixture of Islam and various pagan beliefs that are only found in Egypt's desert areas. The remainder of the Egyptian population are largely Christians—most of them members of the Coptic Church. There are also a few Greek Orthodox, Armenian, Maronite, Anglican, and Protestant Christians, as well as a small Jewish community.

## LANGUAGE

The official language of Egypt is Arabic. Arabic is the language of the Koran. The Koran is the sacred book of Islam and of all Muslims. Arabic is spoken in North Africa and throughout most of the Arabian Peninsula, as well as in other parts of the Middle East.

Standard Classical Arabic, used in books and by the media, differs greatly from the Arabic spoken on the streets. Classical Arabic is the same in all Arab-speaking countries, whereas the dialect of Arabic used in everyday speech varies among countries and areas. In Egypt, an everyday dialect known as Egyptian Colloquial Arabic (ECA) is used.

### The National Anthem

The Egyptian national anthem is called "Biladi!" ("My Homeland!"), with words and music composed by S. Darwish. The English translation of the anthem is:

*My homeland, my homeland, my hallowed land,*
*Only to you is my heartfelt love at command,*
*My homeland, my homeland, hallowed land,*
*Only to you is my heartfelt love at command.*

*Mother of the great ancient land,*
*My sacred wish and holy demand,*
*All should love, awe, and cherish thee,*
*Gracious is thy Nile to humanity.*

*No evil hand can harm or do you wrong,*
*So long as your sons are free and strong,*
*My homeland, my homeland,*
*my hallowed land,*
*Only to you is my heartfelt love at command.*

### Other Languages and Dialects

While people from other Arab nations would be able to communicate formally with Egyptians, in an informal context, such as in the markets or stores, the different dialects of Arabic can cause confusion. There is no official written form of Egyptian Colloquial Arabic, which makes it difficult for foreigners to learn and understand the language spoken on the streets.

The Berber people speak their own language in addition to Arabic. And Coptic Christians, too, use their own language for religious ceremonies. It is derived from a mixture of seven ancient Egyptian hieroglyphs and elements of the Greek alphabet.

# Land and Cities

*"He who has not seen Cairo has not seen the world...
she is the mother of the world."*

*The Thousand and One Nights*

Egypt stands at the meeting place between three continents—a part of Africa, it is nevertheless closely linked to Europe and Asia. Greece and its islands lie to the north across the Mediterranean, while the triangle of Egyptian desert known as the Sinai Peninsula is technically part of Asia.

Egypt covers an area of 386,900 square miles (1,002,071 sq. km)—about one and half times the size of the state of Texas— and lies in the northeastern corner of Africa. The country is roughly square in outline. To the north is the Mediterranean Sea; to the east, Israel and the Red Sea; to the south, Sudan; and to the west, Libya. Egypt's southern and western borders are virtually straight and meet at a right angle deep in the Libyan Desert. Its eastern coastline slopes steeply northwestward like the side of a pyramid, until the Red Sea joins the Gulf of Suez and eventually the Suez Canal. This famous canal links the Mediterranean and Red seas and has been an important shipping route between Europe and Asia since the 19th century.

Only 6 percent of Egypt's land is fertile, most of which lies in the Nile Valley and Delta (the area where the river splits into many parts). For ancient Egyptians, the fertile land around the Nile was all that existed of Egypt. They compared their land to a lotus, with the Nile Delta as the flower and the river as its long, curving stem.

***The minarets (prayer towers) of mosques loom above the rooftops of Cairo; from these the Muslim faithful are called to prayer.***

FACT FILE

- Egypt's highest point is Mount Catherine (Jebel Katrina) in the Sinai Peninsula, which rises to 8,688 feet (2,642 m). At the Qattara Depression in northwest Egypt, the land falls to 435 feet (133 m) below sea level.
- Egypt's Nile River is the world's longest. It rises in Lake Victoria in central Africa and flows northward for 4,132 miles (6,611 km) to the Mediterranean Sea.
- Lake Nasser in southern Egypt was created by the building of the Aswan High Dam in the 1960s. It is the world's largest artificial lake by area, and third-largest by volume.

## REGIONS AND TERRAIN

**The area to either side of the Nile Valley is desert; toward the coast of the Red Sea are mountains. The area where the Nile runs into the Mediterranean Sea (the Nile Delta) is subject to frequent flooding.**

Geographers usually divide Egypt into four broad regions, each with their own special character. These comprise the green, fertile strip of the Nile Valley and Delta, running north–south through the eastern half of the country; the Western Desert, a vast, arid plateau covered with sand dunes and dotted with oases; the Eastern Desert, strewn with wadis—dry river beds—and rising to a range of mountains fringing the Red Sea coast; and the Sinai Peninsula.

### The Nile Valley and Delta

For the last 750 miles (1,200 km) of its 4,132-mile (6,611-km) course, the Nile flows through Egypt, from Lake Nasser on the border with Sudan to the Mediterranean coast between the cities of Mersa Matruh and Rafah. For most of this length, the river runs within a narrow valley, averaging about 10 to

## EGYPT'S TERRAIN

**Fertile valley**
The land of the Nile Valley is regularly flooded and rich in the fertile silts (muds) that wash down from the river's origins.

**Land below sea level**
The vast Qattara Depression in the west of the country is uninhabited. Other areas of low-lying land are around Alexandria and south of Cairo. Generally the height of the land in Egypt decreases as it nears the coast.

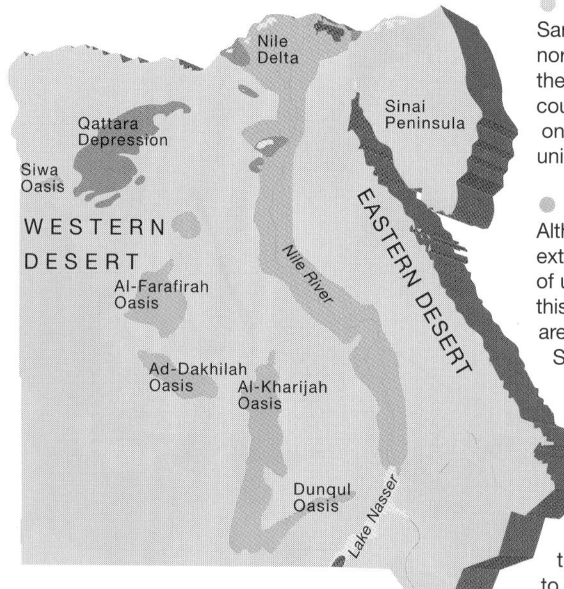


**Sand**
Sandy areas in Egypt lie in the north of the Sinai Peninsula and the south and southwest of the country. The latter region is one of the most remote and uninhabited areas in the world.

**Oases**
Although the Western Desert is extremely arid, it has reserves of underground water. Where this nears the surface, oases are formed, such as that at Siwa in the far west of the country. The southern oases have been part of a recent resettlement program.

**Rocky desert**
The basis of Egypt's terrain is rock. In places this has been eroded to create sand dunes.

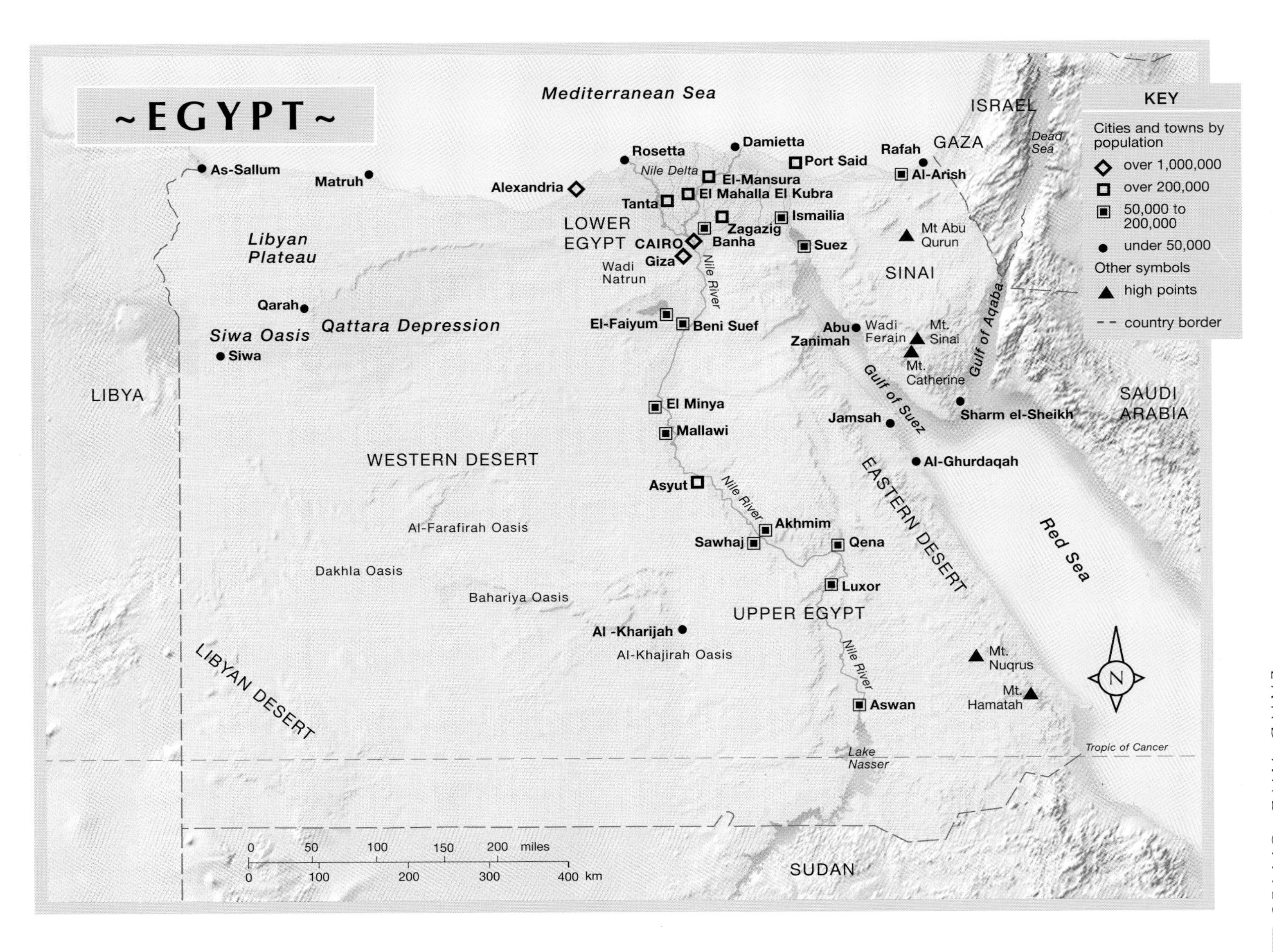
~ EGYPT ~
KEY
Cities and towns by population
over 1,000,000
over 200,000
50,000 to 200,000
under 50,000
Other symbols
high points
country border
Mediterranean Sea
Red Sea
Gulf of Suez
Gulf of Aqaba
Dead Sea
ISRAEL
GAZA
SAUDI ARABIA
LIBYA
SUDAN
SINAI
LOWER EGYPT
UPPER EGYPT
EASTERN DESERT
WESTERN DESERT
LIBYAN DESERT
Libyan Plateau
Qattara Depression
Siwa Oasis
Al-Farafirah Oasis
Bahariya Oasis
Dakhla Oasis
Al-Khajirah Oasis
Wadi Natrun
Nile Delta
Nile River
Lake Nasser
Tropic of Cancer
As-Sallum
Matruh
Qarah
Siwa
Alexandria
Rosetta
Damietta
Port Said
Tanta
CAIRO
Giza
El Mahalla El Kubra
El-Mansura
Banha
Zagazig
Ismailia
Suez
Rafah
Al-Arish
El-Faiyum
Beni Suef
El Minya
Mallawi
Asyut
Sawhaj
Akhmim
Qena
Luxor
Aswan
Al -Kharijah
Abu Zanimah
Wadi Ferain
Jamsah
Al-Ghurdaqah
Sharm el-Sheikh
Mt Abu Qurun
Mt. Sinai
Mt. Catherine
Mt. Nuqrus
Mt. Hamatah
N
0
50
100
150
200 miles
0
100
200
300
400 km

# The Nile River

The longest river in the world, the Nile flows from its origins deep in central Africa to the Mediterranean Sea. Its waters are drawn from no less than eight African countries—Sudan, the Central African Republic, Burundi, Rwanda, Tanzania, Uganda, Kenya, and Ethiopia—and its drainage basin (the area from which its waters come) covers about 1,293,000 square miles (3,349,000 sq. km).

Along its length, the Nile flows through a variety of landscapes, from dense tropical forest and wooded savanna in the south, through swamp and swaying grasslands, to the desert cliffs and rich green fields of Egypt in the north. For about half its length, the river can be traveled by boat. In Egypt, graceful, white-sailed boats, known as *feluccas,* float along the broad river, while large ferries carry passengers from bank to bank. Along the banks are tall palm trees, providing shelter from the harsh midday sun.

For centuries, the source of this great river remained a mystery. Not until the 19th century did European explorers discover its many headwaters. Today the river's prime source is usually taken to be Lake Victoria, whose farthest headwaters rise in the mountains of Burundi. Where the river flows north out of Lake Victoria, it is called the Victoria Nile. After flowing through Lake Albert, it changes its name again, this time to the Albert Nile. As it crosses from Uganda into Sudan, it becomes the Mountain Nile, which then passes into a vast area of swamp and floating vegetation called As-Sudd.

Below the junction with the Sobat tributary, the river becomes known as the White Nile. At the Sudanese capital, Khartoum, the river is joined by its most important tributary, the Blue Nile, which rises in Lake Tana in the Ethiopian Highlands. From Khartoum downstream, the river becomes the Nile proper. At the Sudan-Egypt border, the Nile passes through Lake Nasser (*see* opposite), a vast artificial lake created by the Aswan High Dam (*see* p. 84).

For the last 750 miles (1,200 km) of its course, the Nile flows through Egypt. At Aswan is the island of Elephantine. The island's name was originally *Yebu*, a word that means "elephant." According to some accounts, elephants once bathed on the river banks here, but most likely it was named for the huge gray boulders that dot the southern part of the island and look like bathing elephants.

From Aswan to just south of Cairo, the river is confined in a narrow valley between cliffs of sandstone and limestone. In places these rise to 1,500 feet (457 m). North of Cairo, the river spreads out into the fan-shaped, emerald-green expanse of the Nile Delta.

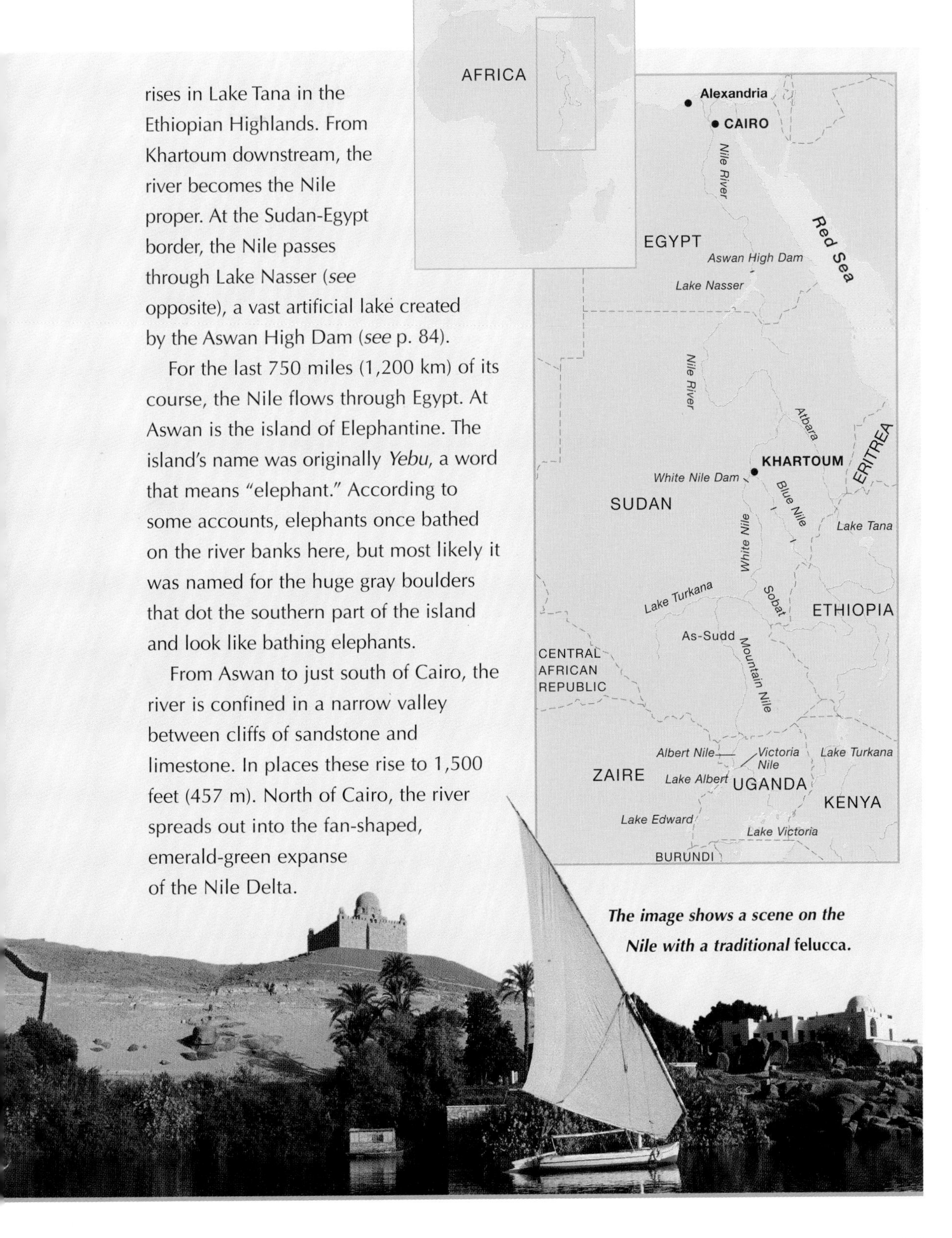


***The image shows a scene on the Nile with a traditional* felucca.**

*Although farming methods have changed little since ancient times, the introduction of new plant hybrids has made agriculture much more productive.*

14 miles (16 to 22 km) wide. The valley is bounded by cliffs rising to desert plateaus on either side.

Near Cairo the cliffs fall away and the river fans out into the maze of channels and canals that is called the Nile Delta. Because the land there is so fertile, this is an intensively cultivated region covering more than 9,500 square miles (24,600 sq. km), an area slightly larger than the state of New Hampshire. It is a uniquely flat landscape that is planted with cotton and rice. Due to the changing course of the Nile, many of the ancient monuments of this region have not survived, unlike those of the drier Upper Nile.

**The Aswan High Dam has increased the amount of irrigated farmland by over a third.**

The Nile carries sediments eroded from the mountains of Ethiopia and, over the years, has deposited silt (particles of soil) on the floor of the Nile Valley to a thickness of 20 to 30 feet (6 to 9 m). This fertile soil provided the basis for the civilization of ancient Egypt and remains the foundation of Egyptian agriculture to this day. The muddy floodplain of the Nile has made it possible for generations of people to live there.

Before the completion of the Aswan High Dam (*see* p. 84), the farmers of the Nile Valley were dependent on the annual flooding of the river to water their fields. The river level was low from January until May, when the waters would begin to rise. The river was fed by the seasonal rains falling on the mountains of Ethiopia and pouring into the river through its tributary (branch of the river), the Blue Nile. Eighty percent of the water that flows through the Nile each year comes from the Ethiopian Highlands. The flood would reach its maximum height in August, and then the waters would begin to recede again, having deposited a fresh layer of fertile mud.

The annual floods were unpredictable, however. In years of drought, the waters did not rise and the crops

***The Nile Delta (below), where the river splits into hundreds of different channels, is ideal for farming due to the rich silt washed down from the mountains.***

**Since the time of the pharaohs, boats known as *feluccas* have sailed on the Nile. Although today few are made of wood, the basic structure of the boats has remained unchanged for centuries.**

would fail. In other years, the floods were too high, washing away fertile soils, houses, roads, and livestock. It was only with the completion of the High Dam at Aswan in 1971 that the mighty Nile was finally brought under control.

## The Western Desert

The fertile strip of the Nile Valley and Delta cuts a green swath through a vast, desert plateau of sandstone and limestone. More than 90 percent of Egypt is desert, and the largest expanse of this lies to the west of the Nile. The Western Desert occupies two-thirds of Egypt's territory—an area the size of Texas—but is

***The ancient town of Shali at the Siwa oasis was founded in the 13th century. Its walls were built of salt, rock, and local clay. Unfortunately three-day rains in 1926 dissolved much of the town and it had to be abandoned.***

## The Siwa Oasis

The Siwa Oasis lies some 350 miles (560 km) west of Cairo, near the Libyan border, and is the most famous and picturesque of the Western Desert oases. It is situated in a depression about 40 feet (12 m) below sea level, where freshwater springs have created a fertile area about 6 miles (9.6 km) long and 4 to 5 miles (6.4 to 8 km) wide.

Siwa is famous for its dates and olives, which have been grown there for more than 2,000 years. There are about 300,000 date palms and 70,000 olive trees. Some 300 freshwater springs are channeled through gardens of palm trees. The houses of the main town cluster around the now-abandoned fortress of Shali, with its ancient mud-brick minaret, or prayer tower (*see* opposite).

Siwa Oasis has been inhabited for thousands of years. It was the site of an ancient Egyptian temple to the god Amun, which was built in the sixth century B.C. Siwa's most famous visitor was the Macedonian ruler Alexander the Great, who came here in 331 B.C. to give thanks to Amun for the successful conquest of Egypt (*see* p. 60).

Having resisted being part of Egypt until the 20th century, the native Berber people are fiercely independent and preserve many of their ancient customs. The first asphalt road in the area was built out of the oasis in the 1980s and a second to a neighboring oasis a few years later. The traditions and way of life at the oasis are now under threat from the growth of tourism.

*The oases of the Western Desert have been inhabited for thousands of years. In the time of the pharaohs and Romans, they formed busy trade hubs on the routes between Africa and the Mediterranean.*

home to only 1 percent of its population. Most of this tiny population live in the remote oasis communities of Siwa, Bahariyya (Al-Bahriyah), Farafra (Al-Farafirah), Dakhla (Al-Dakhilah), and Kharga (Al-Kharijah).

Outside these pockets of habitation, the Western Desert is a hostile expanse of sand and naked rock that supports only a few bands of camel herders. Summer temperatures can soar to a baking 120°F (50°C) during the day, and winter nights can drop below freezing. Rainfall is almost unknown. The landscape is a sea of shifting sand dunes, interspersed with areas of bare bedrock and "desert pavement," where the wind has removed the sand to leave a layer of sandblasted stones and boulders. The ancient Egyptians believed that the god Seth, the murdering brother of Osiris, ruled over the desert, and they therefore regarded the region with fear.

Although there is little or no rain in the desert, sandstone beneath the surface contains water that has filtered through the porous rock from the Nile Valley. In natural depressions in the land surface, where the water table comes close to the surface, water bubbles to the surface in springs and can be tapped by wells and holes drilled into the ground. This precious groundwater gives life to the oasis communities of the Western Desert, feeding plants and humans alike.

The western oases have been occupied since the time of the pharaohs, but recently their populations have swelled as the Egyptian government has introduced land improvement projects and encouraged farmers from the overcrowded Nile Valley to settle in the desert. The oases farms produce dates, olives, wheat, rice, grapes, and citrus fruits.

The Qattara Depression, in the northern part of the desert, is a natural basin that dips to 435 feet (133 m) below sea level. Here, the groundwater that seeps to the surface forms salt lakes and marshes, which make the land impossible to cultivate.

### The Eastern Desert

The Eastern Desert lies between the Nile Valley and the Red Sea and covers about a quarter of Egypt's land area. The region is more mountainous than the

## The Red Sea

The Red Sea, which lies to the east of Egypt, is both an essential trade route and a diver's paradise. Its position, linking Asia to Europe by the much-disputed Suez Canal, ensures that it is one of the busiest waterways in the world.

The sea's huge underwater mountains of coral, shallow reefs that swarm with over 1,000 species of marine life, draw divers from around the world.

No rivers flow into the sea and the limited influx of water from the Indian Ocean—to which the Red Sea is linked by the Arabian Sea—make the water extremely salty. No one is sure where its name originated. One widely held theory is that a type of algae that forms a reddish-brown scum on the sea's surface during the summer gave it its name.

The Red Sea is about 176,060 square miles (456,000 sq. km) in area and reaches depths of over 9,800 feet (3,000 m). In 1989, scientists and conservationists voted the Red Sea as one of the Seven Underwater Wonders of the World.

*The pharaohs called the Sinai Desert the "Land of Turquoise," for the bright blue mineral that was mined here, along with gold and copper.*

Western Desert, consisting of a high, barren plateau bordered in the east by a range of mountains rising to a height of over 7,000 feet (2,130 m). The plateau is crossed by a series of wadis or valleys (*see* box opposite) that have been eroded by streams fed by rainfall on the eastern mountains.

In the east, the mountains run steeply down to the Red Sea coast, which stretches for 500 miles (800 km) from Egypt's northern port of Suez to the Sudanese border. This coast is fringed by spectacular coral reefs, which attract large numbers of foreign tourists and scuba divers. The main resort town is Al-Ghurdaqah, which lies at the mouth of the Gulf of Suez. This strip of coast is now being developed with hotels and resorts for Egypt's tourist industry.

## Sinai

The Sinai Peninsula, which lies in northeastern Egypt, forms a link between the continents of Africa and Asia. It is bounded on the west by the Suez Canal and the Gulf of Suez, on the east by Israel and the Gulf of Aqaba, on the south by the Red Sea, and on the north by the Mediterranean Sea. Its southern part is rugged and mountainous, rising to the peaks of Mount Catherine (8,688 feet; 2,642 m) and Mount Sinai (7,495 feet; 2,285 m), which lie close together. The northern part is occupied by barren desert. Most of Sinai is desert, except for the northern coast where conditions are less extreme.

Water is precious in the desert, and Sinai is famous for its springs and wells, many of which have been used by the bedouin and other desert travelers for thousands of years. Some, such as the Wadi Ferain in the mountains of southwest Sinai, are lush oases thick with date palms. Others, such as the hot springs of Oyun Feroun, which lie north of the town of Abu Zanimah on the west coast, are used for hydrotherapy (bathing as a treatment for ailments).

The population of Sinai is concentrated in the northern town of Al-Arish and in the coastal towns of the southeast. The interior is inhabited by a few bands of bedouin herdsmen. The region is rich in natural resources, including petroleum, coal, gypsum, and manganese; it was in this region that the ancient Egyptians mined copper and turquoise.

The Egyptian government is currently developing Sinai (*see* p. 92) by improving roads, building pipelines to carry imported freshwater from the Suez Canal, and installing desalination plants, which remove salt from seawater, in coastal towns. The idea is to move the landless population of the Nile Valley into Sinai to reduce overcrowding. A similar—largely unsuccessful—policy was carried out in the Western Desert in the 1980s. Tourism is being encouraged, especially on the southeast coast around the resort of Sharm el-Sheikh, which lies at the southern tip of the Sinai peninsula.

## Wadis

A wadi is a dry channel in a desert area that is prone to flash flooding. These valleys may be filled with temporary rivers or streams that support a variety of vegetation and animal life. Due to the sudden flooding and drying effects, the channels may carry large flows of mud or sediment that bring nutrients or seeds, but these may also be carried away again. Important wadis in Egypt include the Wadi Ferain in southwest Sinai, a large oasis, and the Wadi Natrun, a valley to the northwest of Cairo, where salt lakes dry up every summer. Their deposits are used by the chemical industry.

## ADMINISTRATIVE DIVISIONS

**Egypt's centralized economy means that power is effectively concentrated with the government at Cairo.**

The center of political power in Egypt lies in the capital, Cairo. The country is, however, divided into 26 governorates. Each governorate has a governor appointed by the president and is broken down into districts and villages, also run by appointed officials. These appointed officials are assisted at local level by councils elected by the people. The distribution of the Egyptian population is revealed by the cluster of administrative centers around the Nile Delta and the vast governorates of the deserts. In addition to the administrative regions, Egypt is also split into six geographical regions with particular characteristics of terrain. These are the Nile Delta; the Nile Valley from Cairo to Aswan; the Nubian Valley (now filled by Lake Nasser); the Eastern Desert and the Red Sea coast; Sinai; and the Western Desert.

## GOVERNORATES OF EGYPT

Egypt is divided into 26 governorates. They are listed below, followed by their capitals, indicated on the map by a dot.

1. ASWAN Aswan
2. ASYUT Asyut
3. RED SEA Al-Ghurdaqah
4. BENI SUEF Beni Suef
5. BEHEIRA Damanhur
6. PORT SAID Port Said
7. DAQAHLIYAH El-Mansurah
8. DAMIETTA Damietta
9. EL FAIYUM El-Faiyum
10. GHARBIYA Tanta
11. ALEXANDRIA Alexandria
12. ISMAILIYAH Ismailia
13. JANUB SINA El-Tur
14. GIZA Giza
15. KAFR ASH-SHAYKH Kafr ash-Shaykh
16. MATRUH Matruh
17. MINUFIYAH Shibin al-Kawm
18. MINYAH El-Minyah
19. CAIRO Cairo
20. QALYUBIYAH Benha
21. QENA Qena
22. SAWHAJ Sawhaj
23. SHAMAL SINA' Al-Arish
24. SHARQIYAH Zagazig
25. SUEZ Suez
26. NEW VALLEY Al-Kharijah

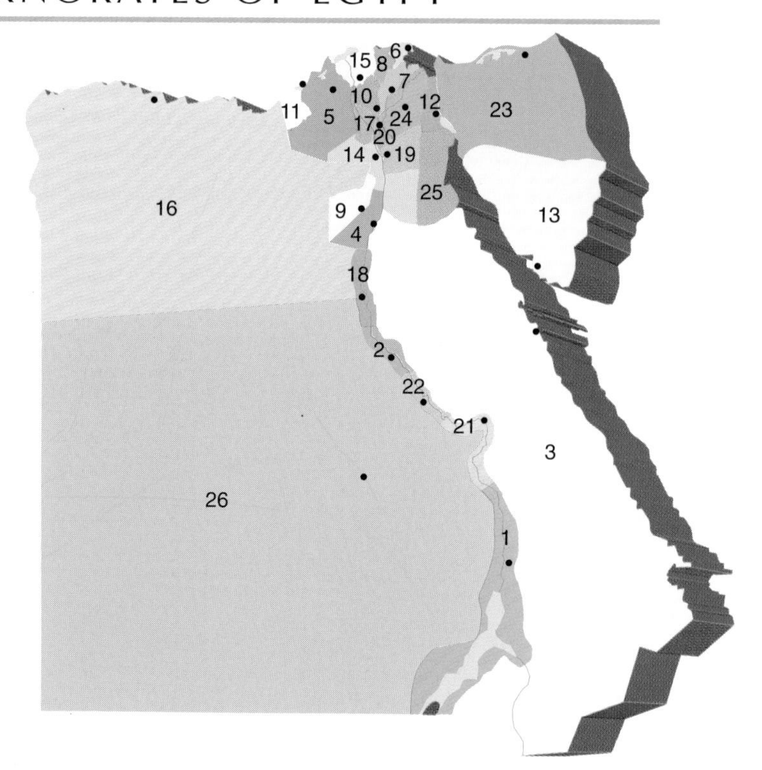

## EGYPT'S CLIMATE

Egypt has a desert climate, and most of the country is hot and dry all year round. The Mediterranean coast between Alexandria and Port Said has a more moderate climate, with typical daytime temperatures ranging from 55°F (13°C) in winter to 80°F (27°C) in summer and an annual rainfall of about 7½ inches (19 cm), which falls from October to April. Everywhere, though, the sun can be very fierce. It is no wonder that the ancient Egyptians thought of the sun god Ra as both creator and destroyer of life.

Rainfall decreases farther south, as the desert climate becomes more extreme. Cairo gets about one inch (2.5 cm) of rain a year. Aswan is lucky to see a half inch (1.2 cm) of rain in five years. The average daily maximum temperature in June is 107°F (42°C). Temperatures in the Western Desert, which are not moderated by the Nile River or by the sea, experience both scorching summers and freezing winter nights.

### The Khamsin

A particularly unpleasant feature of Egypt's climate is the khamsin, a hot southerly wind that occurs between March and May. The khamsin blows in from the Western Desert at speeds of up to 90 miles per hour (145 km/h) and is often accompanied by dust storms. The dust turns the daytime sky deep orange, and Egyptians rush into their homes to shut all doors and windows. No matter how tightly sealed a house is, however, every piece of furniture gets covered with gray dust.

***Very little rain falls south of the city of Cairo. The region of Upper Egypt, around Aswan, experiences scorching temperatures in midsummer.***

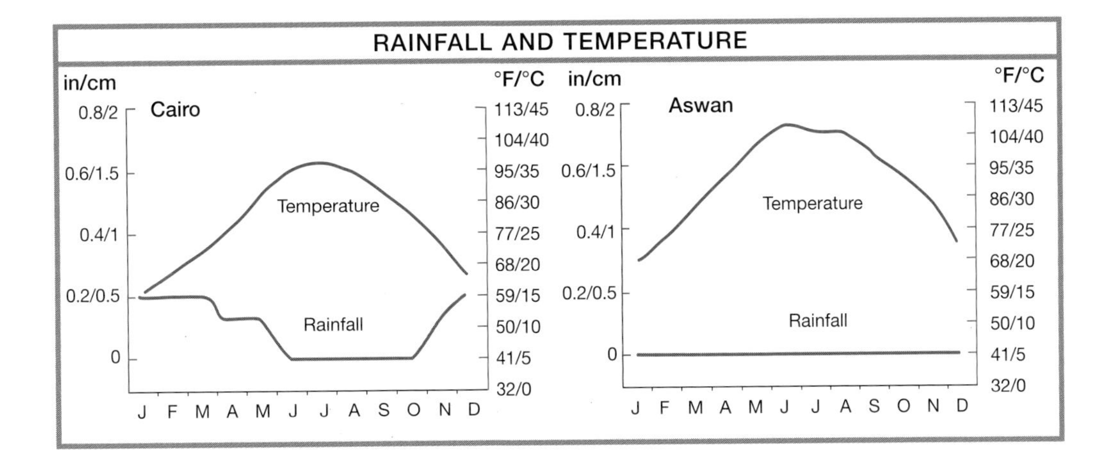

*The fennec, the world's smallest fox, is now extremely rare in Egypt.*

## PLANTS AND WILDLIFE

Egypt's arid deserts support a surprising variety of plants and animals. The lush, fertile lands of the Nile Valley, by contrast, have been farmed for so many thousands of years that many of the original species have died out or have been replaced by domestic animals and plant species introduced from elsewhere.

### Wildlife of the Desert

Large areas of the Egyptian desert—notably the large seas of shifting sand—support no plant life at all. But wherever the water table (the level of water below ground) approaches the surface, in oases or in dry river beds, many species of hardy grasses survive. Tamarisk bushes and acacia trees send down deep roots to tap the precious liquid. These plants may support a variety of animal life.

Many of the large mammals that once inhabited the Egyptian desert have been hunted to near extinction. Species such as the leopard, cheetah, oryx (a type of antelope), and hyena were once widespread, but today they are rarely seen. More common, though still rare, are the Dorcas gazelle, the fennec (a miniature desert fox, no larger than a domestic cat), the Egyptian jackal (*see* p. 30), and the Nubian ibex (another type of antelope).

The rock hyrax, a small mammal like a prairie dog, is found in the mountains of Sinai. Strangely, its closest relative in the animal kingdom is the elephant. Like its massive relative, it is very sociable and lives in large packs. Wild mountain sheep and goats roam the hills and wadis of the Eastern Desert. Lizards, snakes, and scorpions are common throughout the desert regions.

**Camels are thought to have been brought to Egypt by the Persians in the sixth century B.C. They were the main source of transportation until the arrival of railroads in the 19th century.**

## Flora and Fauna of the Nile

One of the most common sights in Egypt is the date palm, which is plentiful in the Nile Valley and in the cultivated oases of the Western Desert. Sycamore and eucalyptus trees, along with casuarina, jacaranda, and scarlet-flowered poinciana—the latter three species introduced from abroad—also grow in and around cultivated areas. In summer, the countryside is bright with their colors.

In the swampy margins of the Nile and in Egypt's thousands of miles of irrigation canals grow bamboo, reeds, and water hyacinth. The Egyptian lotus is a kind of water lily that frequently appears in ancient Egyptian art and is still found in the Nile Delta. Another ancient Egyptian plant, papyrus (*see* p. 31), has vanished from the wild and is now confined to botanical gardens.

The Egyptian cobra can grow to more than 6 feet (2 m) long. In ancient Egypt, the

***The Egyptian cobra is common in the Nile Valley. It feeds on frogs and small birds.***

## The Jackal

Jackals are wild dogs that live in Africa, Asia, and parts of Europe. In Egypt, the mournful cry of a jackal can often be heard at night, echoing across the cliffs of the Nile Valley. For this reason Arab peoples call the jackal, "the howler." The jackal is a nocturnal (night) scavenger, which means that they survive by eating animals that are already dead. The ancient Egyptians associated the jackal with death and the afterlife. They portrayed the god of cemeteries, Anubis (*see* above), with a jackal's head.

pharaohs wore a headdress incorporating the head of a cobra. The striped Egyptian mongoose is a small predatory (hunting) mammal that often attacks and kills cobras and other poisonous snakes. The Nile crocodile, which can grow up to 30 feet (9 m) in length, was once common the length of the Nile, but is now found in Egypt only in Lake Nasser. The last crocodile to be seen north of Aswan was allegedly shot by a British army officer in 1891.

The Nile is home to almost 200 varieties of fish, the most common being catfish and bulti, a spiny-finned fish that is caught in large quantities. The Nile perch grows up to 6 feet (2 m) long and 300 pounds (140 kg) in weight.

**The Nile Valley and the oases of the Western Desert provide a crucial stopover for migrating birds flying toward southern Africa.**

### Wildlife of the Sky

More than 400 species of bird have been recorded in Egypt, although only about one-third of these actually breed there. The country is an important migration route for up to two million large birds a year. The most common local species are the hooded crow, the house sparrow, and the black kite, which are found in towns and villages. The hoopoe, named for its distinctive call,

## Papyrus

Papyrus is a tall, reedlike aquatic plant that grows in shallow water in river margins and marshes. In ancient times, it was grown in the Nile Delta. Its pithy stem—which grows up to 12 to 15 feet (4 to 5 m) tall—was used to make many useful objects.

The central pith was cut into thin strips. These were laid across each other in a crisscross pattern and pressed under a heavy weight. The plant's sticky sap glued the strips together, and after pressing, the papyrus sheet was beaten to make it smooth and pliable.

The resulting material could be used to make cloth, sails, and, most importantly, paper. Papyrus paper was invented by the ancient Egyptians around 2400 B.C. and was used as a writing material for more than 3,000 years. The ancient Greeks and Romans also used papyrus. It was finally replaced by parchment, which was made from animal skins.

Papyrus stems were also used whole, tied together in bunches, to make rafts and boats. Pictures of such boats can be seen in paintings that decorate the walls of ancient Egyptian tombs and temples. In 1969, the Norwegian explorer Thor Heyerdahl (born 1914) built a replica of an ancient Egyptian papyrus boat, calling it *Ra* after the Egyptian sun god. He then attempted to sail it across the Atlantic Ocean from Morocco to Central America. He was trying to prove that ancient Egyptians could have sailed to the Americas. The first attempt failed when the craft foundered in mid-ocean, but the following year he succeeded in sailing *Ra II* to Barbados in the Caribbean.

is common in city gardens and country fields. Herons and egrets fly to the reed beds of the Nile, but the ibis, sacred to the ancient Egyptians, is now found only in Sudan. The lakes and lagoons of northern Sinai attract great flocks of pelicans, spoonbills, and flamingos, while the deserts provide a hunting ground for lammergeier falcons, lanner falcons, kestrels, and golden eagles.

## Egypt's Historical Sites

Upper and Lower Egypt were probably united in about 3100 B.C. with the capital at Memphis, near modern Cairo. However, in the Middle and New Periods the capital moved to Thebes (modern-day Luxor). These two centers contain the bulk of ancient Egypt's monuments, although there have been important finds the length of the Nile Valley. Egypt's hot, dry climate has been a major factor in preserving the monuments.



▲ Once a great center of learning, very little of the original ancient Greek city remains. Recent underwater excavations have revealed important finds in the bay.

▲ Memphis, the ancient capital of Egypt, founded around 3100 B.C., is largely destroyed, although much of its necropolis, Saqqara, has been uncovered. Nearer Cairo lie Egypt's greatest monuments, the famous pyramids at Giza.

▲ The great temple here, carved out of the mountain, was built by Ramses II in the 13th century B.C. Four giant statues of the godlike pharaoh look out across the desert. The other complex built by Ramses at Abu Simbel is the Temple of Hathor. It includes four statues of the pharaoh and two of his wife, Nefertari.

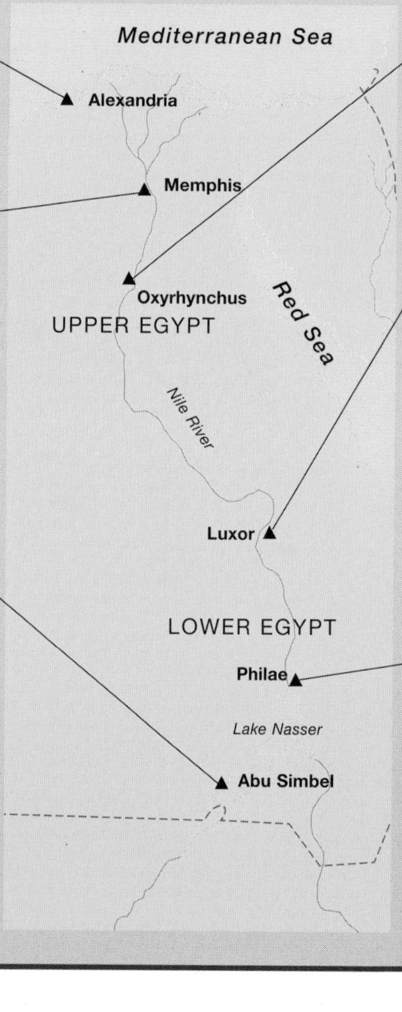

▲ Numerous papyri were found at this regional capital in the 19th and 20th centuries. It was once the third-largest city in Egypt; the papyri were mainly Greek and Roman texts.

▲ With the collapse of the Old Kingdom, Egypt's capital moved from Memphis to the ancient city of Thebes (now Luxor). The center of Egypt's administration for over 500 years, the site contains many of the greatest monuments of the ancient world.

▲ The temple to Isis on the island of Philae was mostly built during the Roman period. When the Aswan High Dam was built in the 1960s, the temple was flooded and the island submerged. The temple had to be removed, stone by stone, to nearby Agilika Island.

## EGYPT'S CITIES

About half of Egypt's population live in the cities that line the Nile Valley and the river's delta region. There was large-scale migration to the cities in the 1970s and 1980s but this has now slowed as overcrowding has begun to make conditions difficult. The main urban employment is in Egypt's vast bureaucracy or in service industries that support tourism. Recently the government has made efforts to encourage light industry.

**Neither Cairo nor Alexandria existed in the times of the first pharaohs. Both were built by later conquerors.**

### Cairo

Cairo, the capital of Egypt, is the biggest city in Africa, and one of the most colorful, crowded, and lively places in the world. In Arabic, the city is known as *al-Qahirah*, from which it gets its Western name. A powerful

*Cut through by the Nile River, the city of Cairo has been built and rebuilt for the last thousand years.*

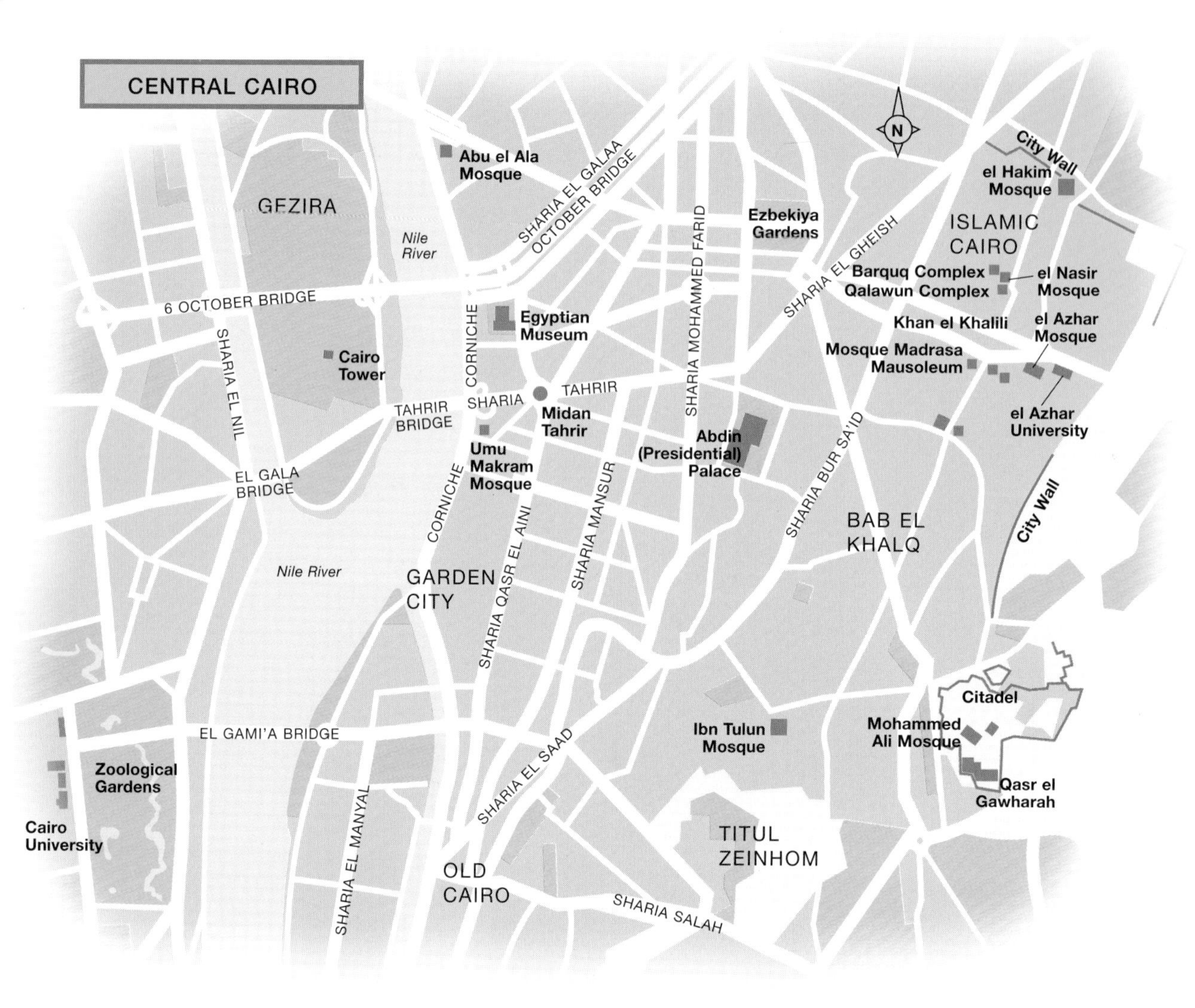


*Central Cairo is a noisy area of shops and banks. Many of the city's main streets converge on Midan Tahrir (Liberation Square).*

dynasty of Islamic rulers called the Fatimids founded the city in A.D. 969. They gave the city its name because its foundation stones were first dug when the planet Mars—which the Arabs call *al-Qahirah* ("victorious")—was rising in the night sky. Much of the city that the Fatimids built still stands today.

Cairo grew very quickly, bursting out of its original walls. The city was renowned for its wealth and splendor. Merchants thronged its streets and bazaars; the domes and towers of palaces and mosques rose over the city skyline; beautiful gardens planted with flowers and containing fountains lay behind their high walls.

In the 19th century, the Egyptian ruler Ismail Pasha (1830–1895) built a whole new center for the city, on marshland to the west of the old city, closer to the Nile River. The king invited prestigious European architects to build his new downtown, creating a Western-style city alongside the Islamic one. Since then, Cairo has sprawled in all directions, and today around 9.7 million people live in and around the city. Many inhabitants are poor peasants from the countryside who have moved to the city in search of work and now live in decaying and overcrowded slums.

**Its rapidly expanding population of 9.7 million makes Cairo the 16th most populous city in the world. It has slightly more people than Paris, but slightly less than Delhi.**

The city center created by Ismail is spacious and stylish. Modern, multi-story buildings line the Corniche—the broad highway that runs along the east bank of the Nile. The bustling central square of Midan Tahrir and the main avenue of Sharia Tahrir are lined with busy shops, restaurants, modern hotels, and movie theaters. The Egyptian Museum, founded by French Egyptologist Auguste Mariette in 1858, houses the world's finest collection of Egyptian antiquities, including the treasures of Tutankhamun's tomb (*see* p. 52), and the mummies (*see* p. 59) of the ancient Egyptian pharaohs.

## Lost in the Bazaar

Bazaars are street markets found in many cities throughout the world; some have been in existence for hundreds of years. One of the largest and oldest is the famous Khan al-Khalili, in Islamic Cairo. The bazaar was founded in the late 14th century and is a labyrinth of narrow streets and alleys, with plenty of dead ends. Hundreds of shops and stalls selling gold, copperware, herbs, spices, fruit, flowers, clothes, leather goods, perfumes, antiques, and even magic spells—all the merchandise of the Middle East—make up the bazaar. Cairenes love to come here to gossip, haggle for a bargain, or to drink a cup of strong, sweet coffee in a coffeehouse, or *ahwa*.

***The only example in Africa, the Cairo subway has two operating lines with one under construction and another still in the planning stages. The main line runs for 27 miles (43 km).***

On a large island in the Nile opposite the city center is the Gezira district, an attractive residential suburb with expensive apartments, foreign embassies, art galleries, and shady gardens. The Cairo Tower, built in 1961, soars 607 feet (185 m) above the island. It is a famous city landmark, and the revolving restaurant at the top offers panoramic views of the city.

East of the modern city center lies medieval Cairo, a great maze of narrow, dusty streets seething with crowds of people, cars, donkeys, carts, and bicycles. The smells of spices, grilled meat, and baking bread mingle with exhaust fumes, dust, and the stench of open drains. In the midst of all this rise the elegant domes and minarets of the al-Azhar Mosque, built in the tenth century. It is linked to al-Azhar University, founded in A.D. 970, which is the oldest university in the world and Egypt's major center of Islamic learning.

CAIRO SUBWAY

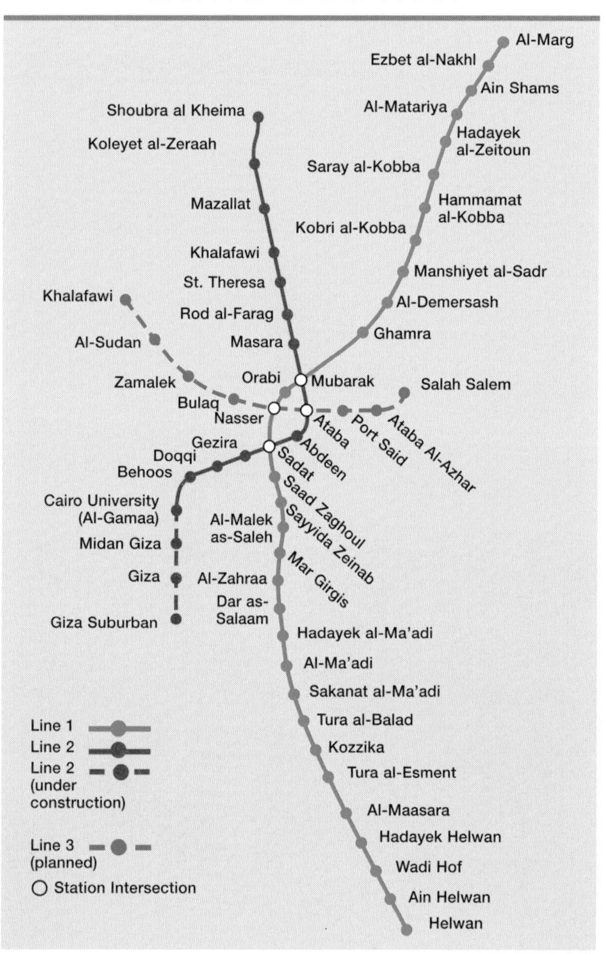


The medieval quarter is overlooked by the Citadel, a fortress founded by the sultan and Muslim hero Saladin (1137–1193, *see* p. 65) in the 12th century, and inhabited by the rulers of Egypt until the 19th century. The walls and towers of the Citadel enclose mosques, palaces, mausoleums, and museums, many built by the 19th-century Egyptian ruler Mohammed Ali. Of the medieval walls that once

## The Nilometer

On an island in the Nile, close to Old Cairo, is the city's famous Nilometer. This was built in the ninth century to measure the rise and fall of the river through the year. The device consists of a column sunk deep beneath the ground and reached by a flight of steps. If the water reached the 16-cubit mark on the column (a cubit was approximately the length of a man's forearm), then the people celebrated because it meant there would be enough water for the crops to flourish.

enclosed the entire district, only fragments remain, including three of the original 60 city gates.

The oldest part of Cairo lies about 3 miles (5 km) south of Midan Tahrir, clustered around a Roman tower built in A.D. 98 by the emperor Trajan. Part of a Roman fortress town called Babylon-in-Egypt, this part of Cairo became a Christian settlement and is still home to several ancient Coptic churches and monasteries. Today, this is one of the most peaceful parts of Cairo, with narrow cobbled alleys and quiet gardens and cemeteries.

***Built from 1830 to 1848, the Mosque of Mohammed Ali is modeled on the Blue Mosque in Istanbul, Turkey.***

The oldest surviving building, the Church of St. Sergius, dates from the tenth century, although an earlier building on the site dated to the fifth century A.D. According to legend, the crypt of the church was built on the spot where the Christian Holy Family rested when fleeing from King Herod after he had ordered the slaughter of male infants. The Coptic Museum contains a magnificent collection of early Christian art and textiles, covering the period A.D. 300 to 1000.

## Giza

The Great Pyramids of Giza (*see* pp. 50–51)—Egypt's most famous tourist attraction—lie just 12 miles (20 km) southwest of Cairo on the Giza Plateau. Formerly a small village, Giza is now the name of a large protectorate on the west bank of the Nile. Once regularly flooded by the river, the area was drained in the 19th century. At the beginning of the 20th century, Giza became home to Cairo University, a counterpart to al-Azhar University in central Cairo.

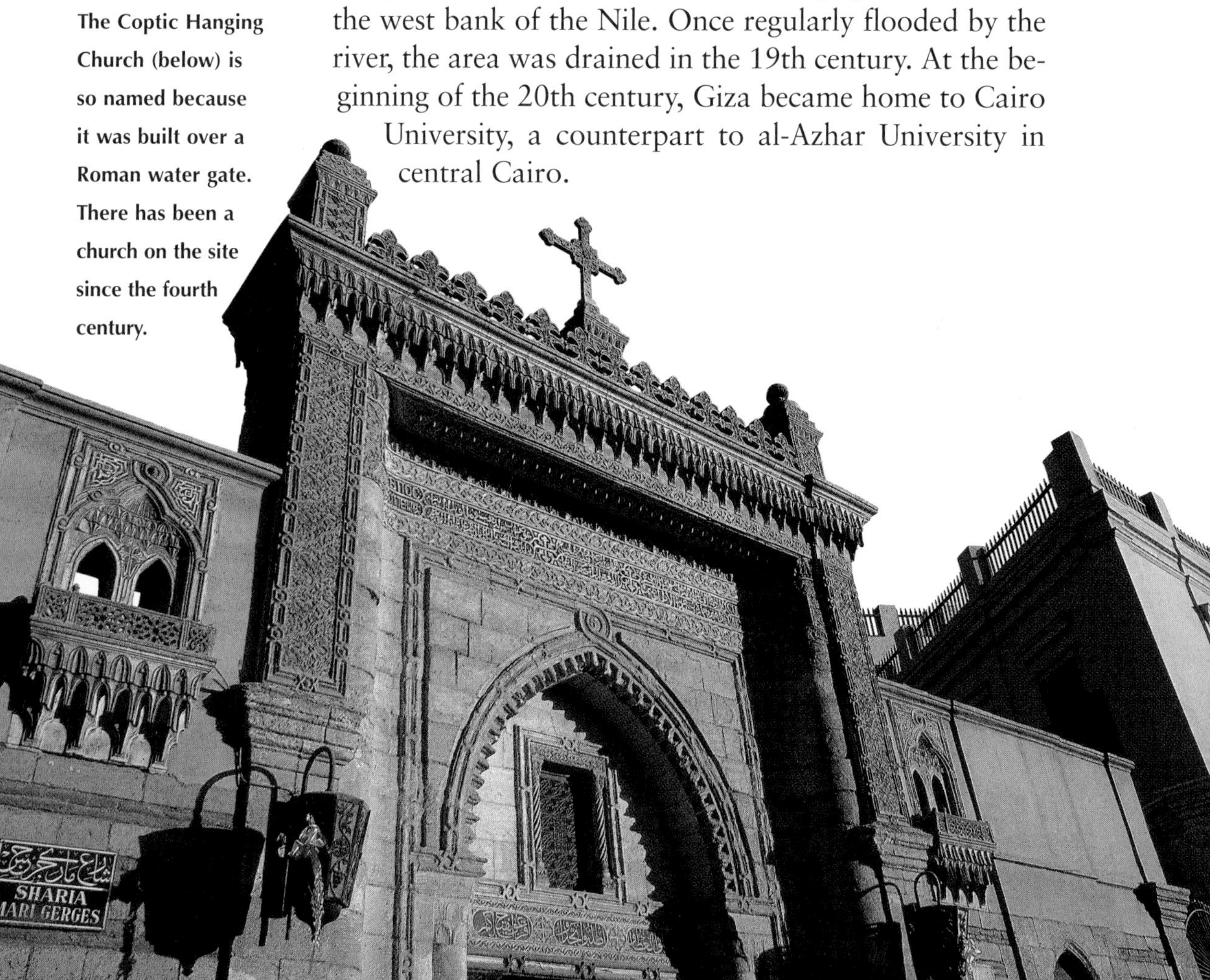


**The Coptic Hanging Church (below) is so named because it was built over a Roman water gate. There has been a church on the site since the fourth century.**

## Alexandria

Located 114 miles (183 km) northwest of Cairo, on the western edge of the Nile Delta, Alexandria is Egypt's main port and second-largest city. The city was founded in 332 B.C. by Alexander the Great (356–323 B.C.), from whom it takes its name. The ancient city was famous for its huge lighthouse—the Pharos—which stood on the island of Pharos in the city harbor. The

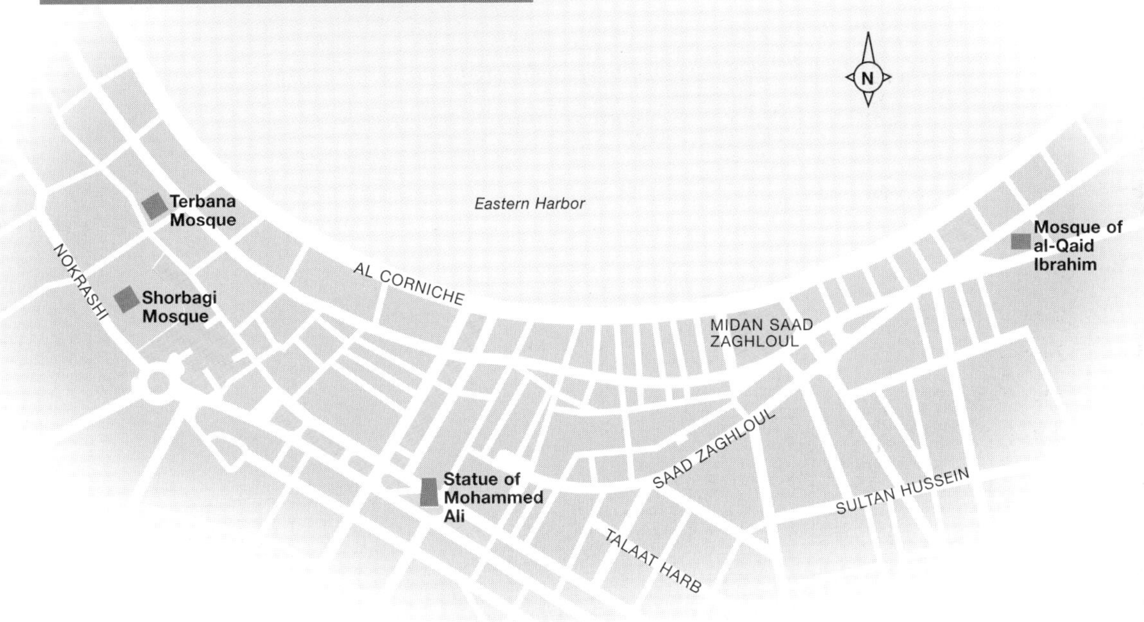


The center of modern Alexandria is based around its curved Eastern Harbor.

lighthouse was over 400 feet (122 m) high and topped by a blazing beacon that guided ships into Alexandria's busy harbor. The ancient Greeks and Romans listed the lighthouse as one of the Wonders of the World. The pyramids at Giza were another. The city was also a center of Greek scholarship and writing. Its massive library was famous throughout the ancient world, but it was destroyed during a civil war in the third century A.D. and its precious books were lost.

Alexandria was the capital of Roman and Byzantine (Eastern Roman) Egypt and was second only to Rome

*Ancient Alexandria was famous for its Pharos or lighthouse. Built by the Romans, it survived 1,700 years before being destroyed in an earthquake. The modern structure above is a fraction the size of the original.*

in size and importance. But it fell into decline after the Arab conquest of 642, and by the end of the 18th century, Alexandria was little more than a fishing village. The city regained its importance in the 19th century after the opening of the Suez Canal.

Modern Alexandria is an attractive waterfront city with a population of 3,700,000. In addition to being a major port and industrial center, Alexandria is also Egypt's most popular holiday resort, with beautiful Mediterranean beaches and fine restaurants. The city stretches for 12 miles (20 km) along the coast, from the industrial Western Harbor to the promontory of Abu Qir, where British admiral Horatio Nelson inflicted a famous defeat on the French emperor Napoleon's fleet in 1798.

Little remains of ancient Alexandria except for the ruins of a Roman theater, the Roman catacombs (the underground burial chambers), and a few columns and gateways, but many ancient objects discovered during the building of the modern city are preserved in the city's Greco-Roman Museum. The site of the Pharos is now occupied by the 15th-century Fort Qait Bey, but recent underwater excavations in the harbor have recovered fragments of the original lighthouse. Twin obelisks known as Cleopatra's Needles once stood at the seaward end of one of ancient Alexandria's main streets. (The name refers to the Egyptian queen, *see* p. 61).They were removed in the 19th century: one now stands on the northern bank of the Thames River in London; the other is in Central Park in New York City.

## Underwater Archaeology

In recent years a joint French-Egyptian team has discovered hundreds of objects off the coast of Alexandria. These include columns, obelisks, statuary, and sphinx bodies. Most impressively, in part of the Eastern Harbor divers have uncovered the remains of a large palace quarter, dating back to the Ptolemaic period (305–30 B.C.). Many of the objects have been returned to the seabed. A Plexiglass tunnel is planned beneath the sea's surface, through which visitors will view an underwater museum.

***Little of ancient Alexandria remains: the city is a mix of colonial and modern buildings.***

## Luxor

*On the east bank of the Nile, ancient Thebes (now Luxor) was the historic religious center in the time of the pharaohs. On the opposite bank are the tombs of the pharaohs, a vast area known as the Valley of the Kings.*

The city of Luxor lies on the east bank of the Nile about 400 miles (640 km) south of Cairo. This attractive riverfront city has a population of 160,000. Although people from the surrounding countryside still come to Luxor to trade their produce, today its principal business is tourism. Luxury hotels line the riverfront, Nile cruise ships moor at the jetties, and the bazaar bustles with tourists and street vendors. The Arabic for Luxor (*al-Uqsur*) means "the forts." These were the palaces that occupied the site of the ancient city of Thebes, a royal capital of ancient Egypt. The city was also adorned with magnificent temples and other buildings. The west bank of the Nile, opposite Thebes, became a vast necropolis (city of the dead), holding the tombs of the pharaohs.

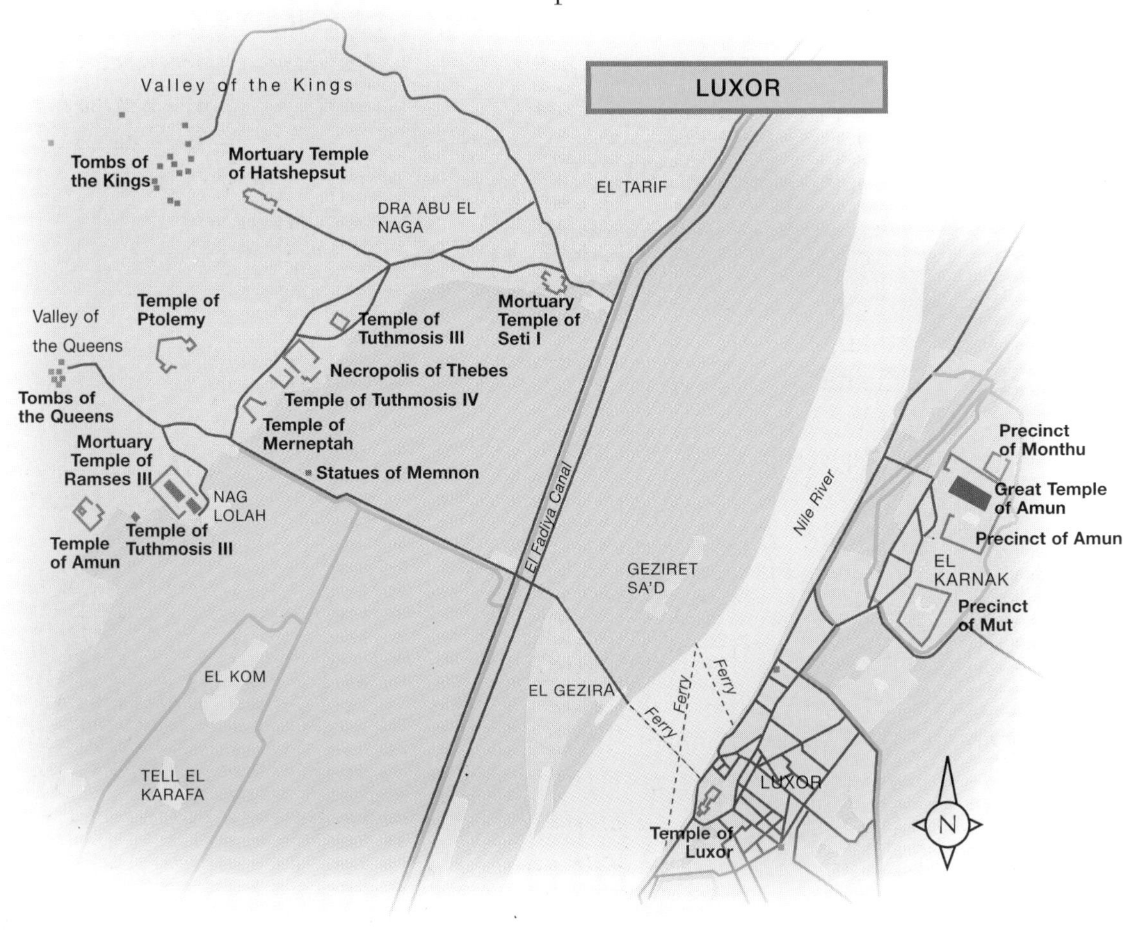

*The processional colonnade of Amenhotep III is at right. On the left stand the pylons of Ramses II at Luxor—huge walls flanked by vast statues of the pharaoh.*

At the center of modern Luxor is the ancient Temple of Luxor, built by Amenhotep III (ruled 1386–1349 B.C.). It was added to by successive pharaohs, as well as by Alexander the Great and Roman leaders. Later, Christian churches were built on the temple grounds, and in the 13th century, a mosque was built in the central court. One of the pair of obelisks that once flanked the main gateway now stands in the middle of the Place de la Concorde, a major square in Paris, France.

Two miles north of Luxor, at Karnak, stand the ruins of the Great Temple of Amun, built over a period of 1,500 years. This was the greatest temple complex in Egypt, and one of the largest in the world. Its most impressive features are the remains of the Great Hypostyle Hall, built by the pharaohs Seti I (1291–1278 B.C.) and Ramses II (1279–1212 B.C.). A forest of 140 towering stone columns covers an area of 5,600 square yards (4,850 sq. m)—about the size of an American football field.

**In 1997 an attack by Muslim extremists killed 58 tourists at the Temple of Hatshepsut on Luxor's west bank. The town's tourist industry has yet to recover.**

A ferry trip across the Nile leads to the necropolis on the west bank. Here, the cliffs in the Valley of the Kings and the Valley of the Queens are full of the secret tombs of pharaohs, queens, and high priests. Huge mortuary (funeral) temples rise on the plains below, where carvings and inscriptions record each monarch's achievements in life and chart their journeys through the underworld (the place of departed souls) after death.

# Past and Present

*"All the world fears Time, but Time fears the Pyramids."*

Egyptian proverb

Egypt has the longest recorded history in the world and is home to one of humankind's oldest civilizations. Set at the point where Africa and Asia meet, the fertile valley of the Nile River has sustained human settlement for at least 7,000 years. The Nile Valley was protected from invaders by the vast expanses of desert on either side of it, enabling the ancient Egyptians to develop a sophisticated society over an extended period of time without the fear of invasion. When invaders did reach ancient Egypt, they were often so impressed by its advanced culture that they preserved and learned from it. As a result, Egypt has one of the best preserved civilizations of the ancient world.

Ancient Egyptian society was very advanced. Its surviving monuments show the enormous power wielded by the pharaohs—the country's rulers in ancient times—as well as the skill of its engineers and artists. Ancient Egypt had a developed system of government, irrigation, and picture-writing (hieroglyphics), and laid down the foundations of modern astronomy, mathematics, and medicine. Modern knowledge about life in ancient Egypt, the study of which is known as Egyptology, has come from the great monuments and tombs that still exist today. Decay or rotting is less likely in a dry climate such as that of Egypt, so a vast number of ancient artifacts have survived.

***The tomb painting opposite shows a scribe offering flowers to the God of the West; it dates from about 1186–1069 B.C. and is now in the Cairo Museum.***

## FACT FILE

- Two of the Seven Wonders of the Ancient World—the Pharos (lighthouse) of Alexandria, and the Great Pyramid of Khufu at Giza—were built in Egypt. Only the Great Pyramid survives.
- The English word "pharaoh" is derived (via Greek and Latin) from a Hebrew word that was used in the Bible to describe the kings of ancient Egypt. It originally came from the ancient Egyptian word *per'aa*, meaning "great palace."
- Ancient Egypt was ruled by families—or dynasties. Traditionally each dynasty is given a number.

## PREDYNASTIC EGYPT

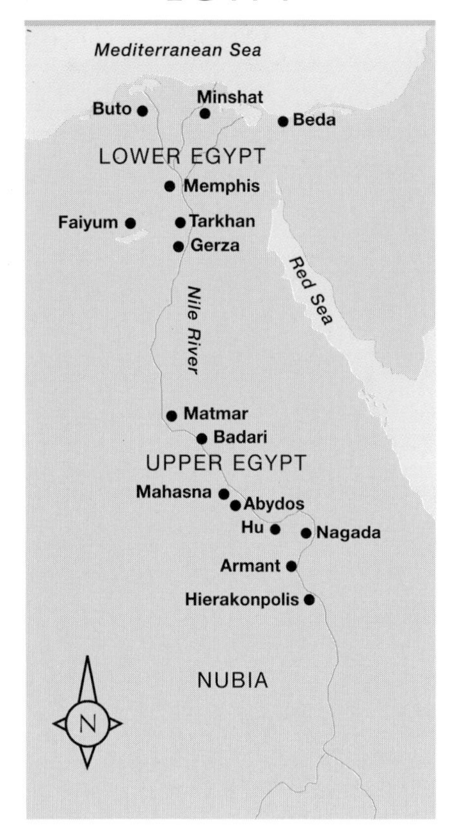

*The symbol of Upper Egypt was the white crown or* hedjet *(top); Lower Egypt was represented by the red crown or* deshret *(middle). Upper and Lower Egypt united were symbolized by the double crown or* shmty *(bottom).*

## Egypt of the Pharaohs

Egypt's ancient civilization flourished for almost 3,000 years before it succumbed at last to foreign invasion. This remarkable early history resulted from Egypt's unique relationship with the Nile. The lush Nile Valley was protected from invasion by huge deserts on either side of it, the waterfalls and swamps of Nubia to the south, and the marshy delta to the north. The rich farmland provided such plentiful harvests that the people were well fed, and the need for irrigation and a unified workforce led to organization and cooperation among the citizens and the creation of the world's first nation-state.

Egypt's recorded history traditionally begins with the unification of the kingdoms of Upper and Lower Egypt under King Menes, around 3100 B.C. Upper Egypt consisted of the Nile Valley from Memphis south to what is now Aswan, while Lower Egypt encompassed the lands of the Nile Delta, to the north. Menes took the title of "Lord of the Two Lands," and founded a new capital, Memphis, on the west bank of the Nile about 15 miles (24 km) south of modern Cairo.

## The Old Kingdom

The Old Kingdom (c.2686–c.2184 B.C.) covers the period of the Third to Sixth Dynasties, when the vast pyramids of ancient Egypt were built. During this time Egypt's wealth expanded due to campaigns in Libya, Nubia, and the Sinai. The organization of the state was a tight one, with most of the major positions of state being held by relatives of the pharaoh. The pharaohs of the Third Dynasty (c.2686–c.2613 B.C.) began the construction of pyramids for their royal tombs. Imhotep, the architect of the Third-Dynasty pharaoh, Zoser, is the first architect in history that we know by name. He developed the simple rectangular tomb chamber, or mastaba, into a series of mastaba built one on top of the other, thus forming a step pyramid (*see* pp. 50–51). During these centuries, the power and prestige of the pharaohs reached new heights. In particular, the Fourth-Dynasty (c.2613–c.2500 B.C.) kings Snefru, Khufu, Khafre, and Menkaure constructed huge pyramids for themselves at El-Faiyum, Saqqara, and Giza. These vast building projects were enormously costly and demanded the forced labor of many thousands of Egyptian subjects, who would have worked on them seasonally.

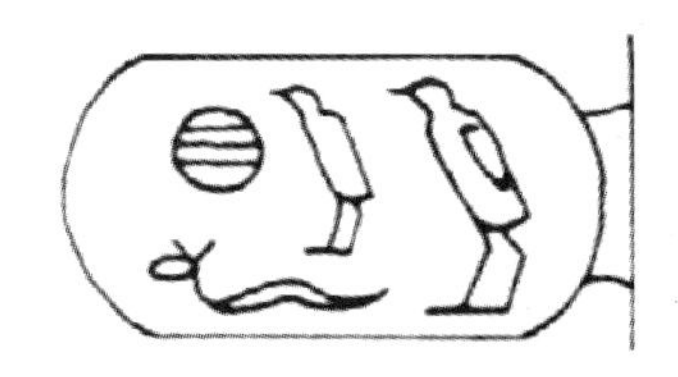

*Seal of Khufu (Cheops) 2589–2566 B.C., Fourth-Dynasty builder of the Great Pyramid at Giza.*

*Keeping records was an essential part of ancient Egypt's successful government. This clay scribe dates to the Fourth Dynasty (c.2613– 2500 B.C.).*

Toward 2350 B.C., the power of the pharaohs began to dwindle and control became increasingly shared with local princes and noble families. The pyramids built during this period were much smaller and generally of poorer construction. The tombs of the nobility were no longer to be found near the center of government, but in out-lying provinces. This suggests the disintegration of the central authority of the pharaoh. Struggles for dominance among the various Egyptian princes led to civil war during the time of the 11th Dynasty (c.2134–1991 B.C.) and the emergence of a new kingdom with its center at Thebes (present-day Luxor) in Upper Egypt.

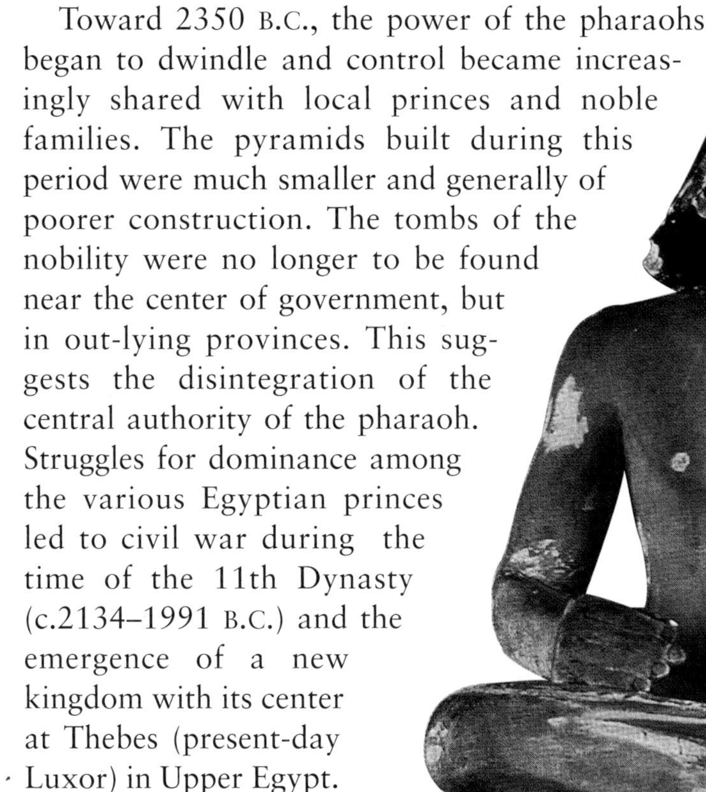

## The Nubians

Nubia was a region to the south of ancient Egypt. It was regarded by the ancient Egyptians as a source of gold and an important trading region for ivory, cattle, and slaves. As a result, between 2000 and 1000 B.C., Nubia was taken over numerous times by Egyptian rulers. Thereafter, Egyptian power declined and Nubia was ruled by the Kushites. From 750 to 650 B.C. the Kushites—Nubia's most important civilization—ruled Egypt and their culture became increasingly Egyptianized. By 650 B.C. the Assyrians had driven Egypt out of the region, and Nubia became more African in character.

### The Middle Kingdom

The protection of the Nile Valley had made the rulers of the Old Kingdom complacent in their great wealth. The breakdown of this centralized order was shocking and a period of chaos followed. The situation was eventually brought under control by a series of strong princes from Upper Egypt, with their center at Thebes. These new rulers managed to reunite the whole country and once again centralize government. The Middle Kingdom (2040–1782 B.C.) was a period of renewed prosperity, which led to an increase in the popularity of the cult of Osiris, Lord of the Afterlife. His spiritual center at Abydos became a site of annual pilgrimage and a place where the pharaohs would erect tombs to ensure their future in the afterlife.

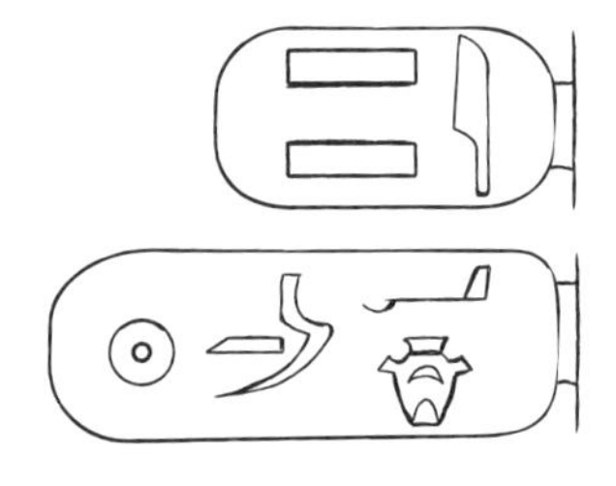

***On the right is the seal of Sheshi, one of the 15th-Dynasty pharaohs, who were known as the Hyksos, or "Desert Princes."***

It was during this time that many fine tombs, temples, and royal palaces were built throughout Egypt. The kings of the 12th Dynasty (1991–1782 B.C.) built pyramids and monuments at El-Faiyum and Saqqara. The first of these, which no longer exists, was described by the Greek historian Herodotus as more impressive than the great pyramid at Giza. The princes of this period also led military campaigns south into Nubia, north into Palestine, and even as far afield as Syria and Greece. Nubia, taken by conquest, became part of Egypt. Its seizure allowed trade links with parts of Asia to the east of Nubia to be established. But the end of the Middle Kingdom was marked by more power struggles among the Egyptian provinces, which left the country vulnerable to invasion.

The Hyksos were invaders from the areas around Palestine or Syria. Known to the Egyptians as "Desert Princes," they captured Lower Egypt and set up their capital at Avaris in the Nile Delta. The newcomers brought with them many innovations, including new metalworking techniques, crops, animals, and weapons. The horse and chariot were the most important of these innovations. The Hyskos were skilled archers and their use of chariots brought advanced warfare to the area. However, by 1500 B.C., the establishment of a secure government at Thebes had enabled the armies of Upper Egypt to drive out the invaders and reunite the two lands once again. The success of this campaign led to the greatest period of ancient Egypt's history—the New Kingdom.

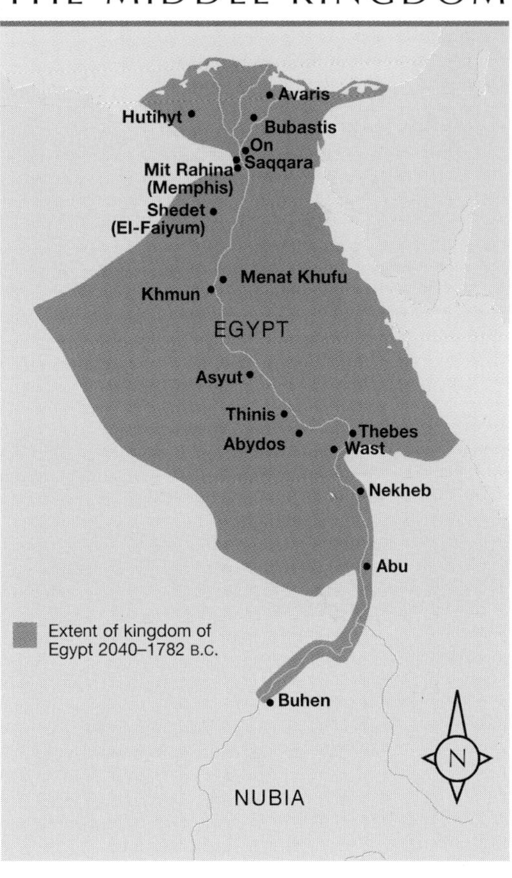


*The map above shows the extent of the kingdom of Egypt during the Middle Kingdom (2040–1782 B.C.).*

# The Pyramids of Egypt

The pyramids of Egypt are probably the most famous ancient monuments in the world. Around 90 pyramids are known, and while many have been reduced to piles of rubble, those that have survived rank among the most impressive feats of architecture in history.

The great age of pyramid building lasted for some 2,500 years, from the beginning of the Old Kingdom to the close of the Ptolemaic Dynasty (30 B.C.). From c.2650 to c.2150 B.C., almost all Egyptian royal tombs took the form of pyramids. Most pyramids of this period were built along the west bank of the Nile, from the site of modern Cairo south to Beni Suef. The oldest is the Step Pyramid of King Zoser, built at Saqqara, 16 miles (25 km) south of Cairo, in the 27th century B.C. Before this, Egyptian kings were buried in tombs made of dried-mud bricks. These mastabas, as they are called, were often in the form of a low, stepped mound, built over an underground burial chamber.

Pyramid building reached a peak during the Fourth Dynasty (c.2613–c.2498 B.C.), when the famous pyramids at Giza were built. The pyramids at Giza were known to the ancient Greeks as one of the Seven Wonders of the World, and today they are one of the world's most popular tourist attractions. The three large pyramids are (above, left to right) the Pyramid of Menkaure, the Pyramid of Khafre, and the (Great) Pyramid of Khufu.

The Great Pyramid of Khufu (built 2600–2500 B.C.), is the largest of all.

Each side measures 755 feet (230 m) at the base, and the pyramid was originally 480 feet (146.5 m) high. The top 30 feet (9 m) are now missing. The sides, like the other pyramids at Giza, are aligned almost exactly north–south and east–west. The Pyramid of Khafre is almost as big, but the Pyramid of Menkaure is much smaller, at 203 feet (62 m) high.

Scholars have not fully explained how these astonishing buildings were built. Their vast size, fine stonework, and precise measurements imply a very advanced knowledge of architecture and technology. The building of the pyramids also involved a high level of social organization and discipline. Farm laborers, who made up 90 percent of the population of ancient Egypt, had to work on the pyramids as a form of taxation. The work was done when the Nile was in flood and when it was therefore impossible to farm the land.

Astronomers figured out the precise angle at which the pyramid should stand. Stone blocks were laid out as terraces to enable the laborers to stand on the sides of the pyramid as they constructed it. After the passageways and chambers inside the pyramid were constructed, the terraces were filled in.

Much of the stone used to build the pyramids came from local quarries and was transported on barges up the Nile. The first stone blocks were probably set in place with simple cranes and levers. As the pyramid grew, ramps were used to haul up stone blocks with ropes. An outer covering of limestone was applied and polished to make it shine.

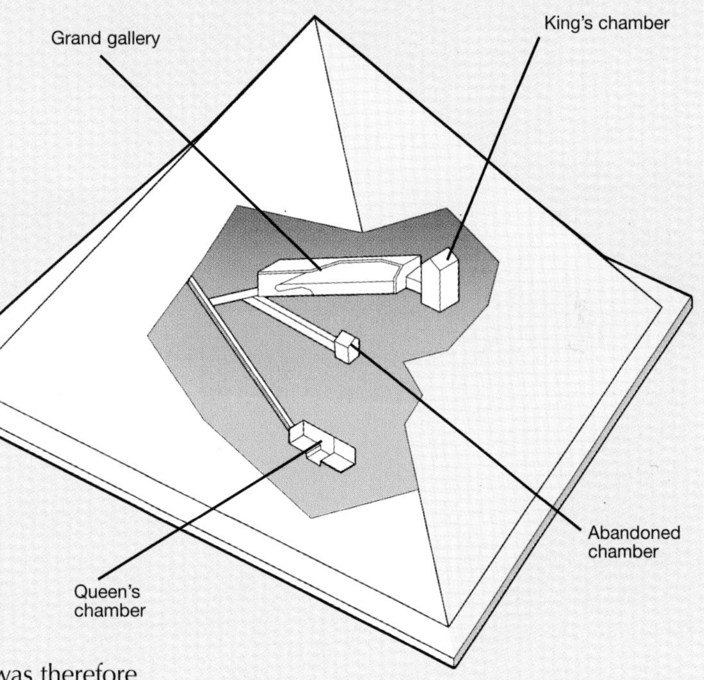


**INTERIOR OF THE GREAT PYRAMID, GIZA**
***The massive exterior of the Great Pyramid belies the small internal galleries where Khufu was buried. An underground chamber was abandoned in favor of an ascending passage cut into layers that had already been completed.***

*Above is the seal of the boy-pharaoh Tutankhamun.*

**The New Kingdom**

Under the rulers of the New Kingdom (c.1570–1070 B.C.), Egypt grew into an international power. Its military campaigns built an empire that spread into western Asia. Prosperity returned as the empire received tribute from subject countries and merchants traded in Phoenicia, Crete, and around the Aegean Sea. Massive temples were built throughout Egypt, most notably the Great Temple of Amun at Karnak, just outside Thebes. Pyramids were abandoned, and pharaohs were instead buried in rock-hewn tombs in the famous Valley of the Kings, west of Thebes.

The tombs were decorated with colorful paintings and inscriptions showing the dead pharaoh's journey through the underworld and filled with goods and treasures that he would need in the afterlife.

## Tomb of Tutankhamun

The most famous burial chamber in the Valley of the Kings is the tomb of Tutankhamun, discovered by English archaeologist Howard Carter in 1922. Tutankhamun inherited the throne at the age of nine, and died around 1325 B.C. when he was only 18. His tomb was remarkable because it had not been broken into by grave robbers and the extraordinary treasures that had been buried with the young king were still intact. They are now displayed in the Egyptian Museum in Cairo.

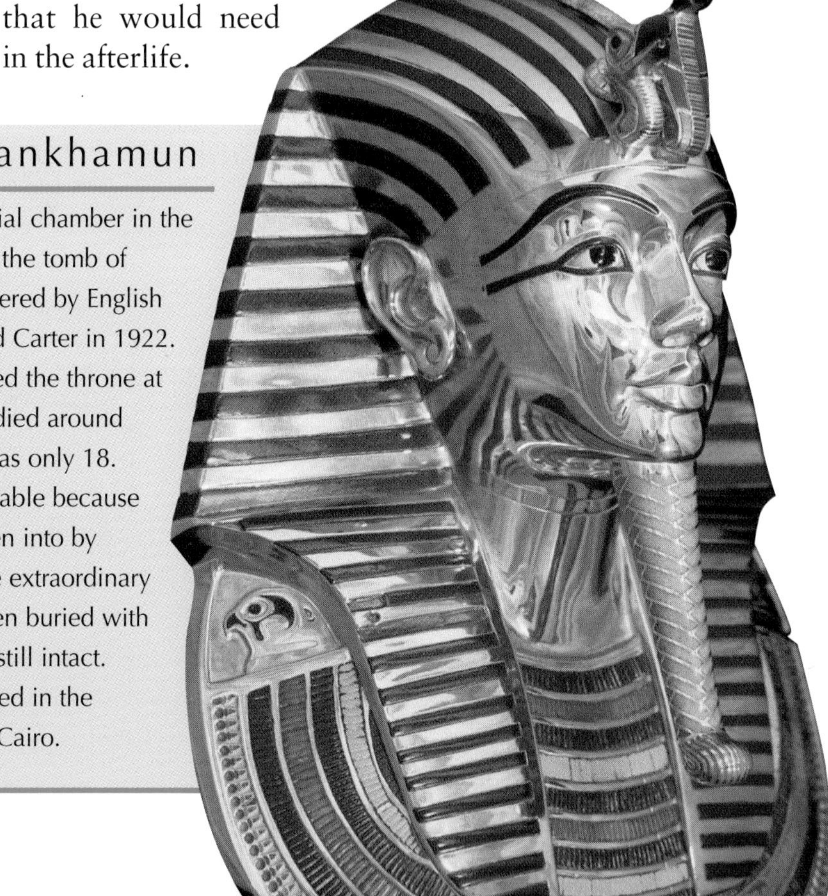

Such tombs were built in secret and were cleverly concealed in an attempt to prevent grave robbers from breaking in and stealing the priceless objects buried along with the pharaoh.

One of the most famous rulers of the New Kingdom was Tuthmosis III (reigned 1504–1450 B.C.), who expanded the Egyptian empire into western Asia beyond Palestine and Syria. Amenhotep III, who reigned from 1390 to 1349 B.C., built many temples, palaces, and monuments, including the Temple of Luxor and his own vast mortuary temple at Thebes. His wealth came not from conquest but from international trade and gold, mined in Upper Egypt. Akhenaten (1350–1334 B.C.) created a religious cult devoted to the worship of a single sun-god, Aten. Perhaps the greatest of the New Kingdom pharaohs, Ramses II (1279–1212 B.C.), signed the Treaty of Kadesh, which was the first recorded peace treaty, with the Hittite king Hattusilis III.

*The map below shows the extent of the Egyptian empire during the New Kingdom (1570–1070 B.C.) and the land it controlled in Syria and Sinai.*

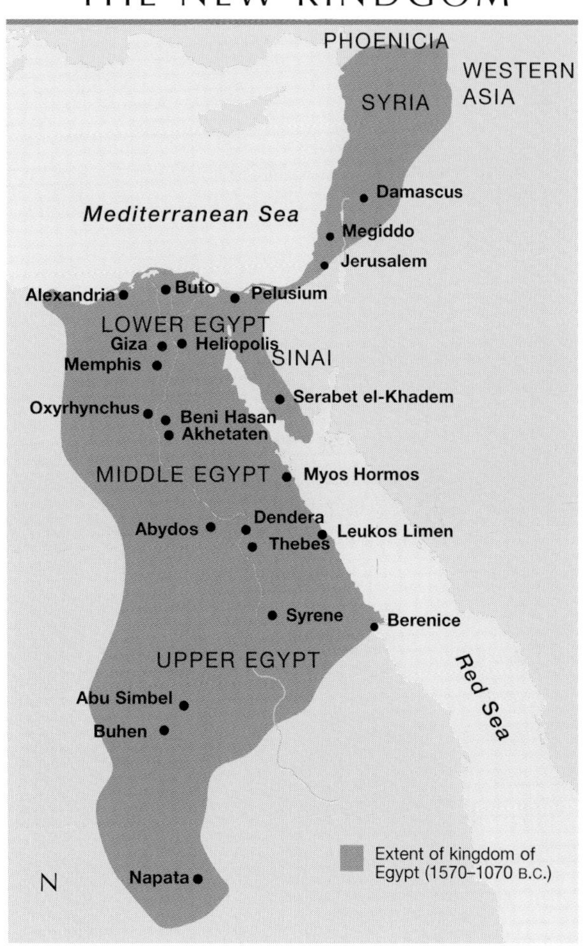


The final century of the New Kingdom was marked by wars, a shrinking empire, a deteriorating economy, and social unrest. Egypt suffered internal strife and invasions from Assyria, Libya, Ethiopia, and Persia. The country had been under Persian rule for almost two centuries when Alexander the Great, a young Macedonian general, invaded in 332 B.C.

## LIFE IN ANCIENT EGYPT

***This tomb painting from the 18th Dynasty (c.1590–1293 B.C.) shows a hunting scene. The figure holds three decoy herons.***

Life in ancient Egypt revolved around the annual Nile flood, which watered the land and laid down a fresh layer of fertile silt. The main crops were wheat and barley, and the Egyptians' diet was varied with fruit and vegetables and fish from the river. Cattle were kept for milk and used as draft animals to pull the farmers' plows. Meat was eaten only by the wealthy.

Papyrus reeds (*see* p.31), which grew in the marshes of the Nile Delta, were tied together in bunches to make rafts and boats or woven into matting. Papyrus was flattened to produce paperlike sheets for writing. The river also provided a means of transportation that united all parts of the kingdom. The prevailing wind in Egypt blows from the north, so boats and barges could make their way upstream under sail and return by drifting down on the current.

### How Ancient Egypt Was Governed

**The Egyptians believed that their pharaoh was a god on earth. When he held the symbols of power, it was thought dangerous to even touch him.**

Ancient Egypt's periods of stability and prosperity were a result of strong central government. People were also brought together by shared religious beliefs.

The pharaoh had total power, made all the laws, and owned most of the land. During the New Kingdom, the country was divided into 42 districts, each of which had its own local administration. The pharaoh employed thousands of advisers and officials to administer them and all aspects of Egyptian life, including the army, land,

## Gods and Goddesses

The ancient Egyptians worshiped hundreds of gods and goddesses. Many were worshiped only locally, but some were recognized as great gods throughout the country. To make matters more complicated, one god could take many forms or be identified with another. The gods were often depicted in Egyptian art and might be shown with an animal head or sometimes entirely in animal form. Below is a list of some of Egypt's most important gods.

**Bastet** The cat goddess, shown in the form of a woman with a cat's head. At her chief temple in the city of Bubastis, joyous festivals were held in her honor.

**Hathor** The goddess of love, who protected women and travelers. She was depicted in the form of a cow or as a beautiful woman with cow's ears.

**Horus** The god of the sky and son of Isis and Osiris.

**Isis** The wife of Osiris and mother of Horus. After Osiris was killed, she restored him to life again. She was depicted as a woman with cow's horns, with a sun disk between.

**Osiris** God of the underworld, who ensured the fertility of the earth. He appeared holding a crook and flail (a tool for threshing grain), which represented his power. He was murdered by his evil brother Seth and his body torn to pieces.

**Ra** The sun-god. The Egyptians believed that every day Ra sailed through the sky in a solar boat before returning to his home in the underworld. In the New Kingdom, his worship merged with that of another powerful god, Amun. As Amun-Ra, the god was considered to be the father of the pharaoh.

**Thoth** The god of writing, counting, and knowledge. He took the form of a baboon or ibis.

and religious temples. All Egyptians had the right of appeal to the pharaoh, who, as the supreme ruler, head of the army, and representative of the gods, was the source and embodiment of all justice.

## Work and Trade

Most of the work done in ancient Egypt, such as farming and pyramid building, was forced labor. Slavery first appeared in Egypt during the New Kingdom (1570–1070 B.C.), when the conquest of other countries enabled the pharaohs to import foreign laborers. The best-known example of slavery in Egypt is probably the enslavement of the Hebrews, who, according to the Bible, had been led by Moses in their exodus (mass departure) from Egypt, probably in the reign of Ramses II (1279–1212 B.C.). The Hebrews entered Egypt during a period of rule by Semitic conquerors called the Hyskos. When the Hyskos were deposed around 1570 B.C., Hebrews were forbidden to leave and became slaves.

**The painting above shows ancient Egyptians picking and treading grapes to make wine.**

Workers in Egypt were mainly farm laborers, some of whom worked for administrators or for the pharaoh. There were many highly skilled craftsmen who produced basic essentials such as bricks, baskets, papyrus, mats, tools, and cooking vessels.

There was a sizeable class of traders and boat-owners, who provided transportation along the Nile. Sailing was an important occupation and essential to the economy of ancient Egypt since the Nile provided the only practical means for transporting goods up and down the country.

Payment was made by barter—the exchange of goods—rather than by money. Much domestic trade probably existed between farmers and urban craftsmen. Foreign trade developed during the New Kingdom, when Egypt expanded beyond its borders. The country exported grain, linen, and papyrus in exchange for timber from Lebanon, copper from Cyprus, and incense from the East. Gemstones came from as far away as Afghanistan.

The ancient Egyptians saw their vast empty desert as an alien area. They ventured there only to obtain precious metals, such as gold, and gemstones. The Nile Valley, on the other hand, was their fertile homeland where the gods ensured that the annual flood came without fail. The Nile flood marked the start of the ancient Egyptian year in July. This coincided with the reappearance in the night sky of the star Sirius. The Egyptians associated Sirius with the goddess Isis, whose tears were believed to cause the flood.

## Farming

The agricultural year was divided into three seasons: the flood, when the waters of the Nile covered the fields; the sowing season; and the harvest. Wheat, barley, and flax for weaving into linen were the main crops. Seeds were sown by hand—usually by the farmer's wife—and wooden plows drawn by oxen were used to tread the seed into the soil. After planting seeds, the farmers would maintain the irrigation channels and attempt to protect the crop from pests. After the predicted yield of the crops had been calculated for tax, they gathered in the wheat. Tax collectors then compared the estimate to the actual yield to ensure no crops had been dishonestly kept back by the farmers.

***This wooden model of a woman grinding grain dates to the Middle Kingdom (2040–1782 B.C.). The stylized form of the woman's body shows the sophistication of even the most simple Egyptian artwork.***

**The shadoof was a hollowed out pole with a bucket at one end and a counterweight at the other. It enabled the farmer to transfer water between levels without bearing the load.**

The management of the country's water supply was everyone's responsibility. While every citizen was dependent on water for growing food, the country's taxation also depended on its water supply. The height of the Nile flood was recorded each year on specially built steps that led down to the river. These enabled state officials to measure the water level and assess the potential of the harvest and thus the rate of taxation (*see* p. 37).

In order to pay their taxes, therefore, farming communities had to use the floodwaters effectively. The water had to be kept from running off the fields when in flood, and from evaporating before it had soaked into the soil. The crops also had to be watered. To do this, the Egyptians developed a network of irrigation canals. A device called a shadoof was also invented (*see* panel, left), which allowed the water to be raised up to about 6 feet (2 m). This was an important innovation that allowed the Egyptians to increase the amount of land under cultivation by about 15 percent.

## Science, Astronomy, and Medicine

The ancient Egyptians' remarkable achievements in science, astronomy, and medicine were often tailored to religious purposes. Astronomy—the study of the movement of the planets and stars—was used mainly for religious observances and for the calendar. Astronomers observed solar and lunar eclipses, and it is thought that they may have seen Halley's comet. The Egyptians were brilliant mathematicians and were particularly skillful in geometry. The pyramids have been shown to be based on subtle geometric harmonies. Alchemy, a type of ancient chemistry, was also practiced.

Findings from tomb decorations and mummies have provided us with a picture of health and disease. Symptoms of ailments have been found listed on papyrus documents, with medicines for their treatment.

*The symbols shown below are known as glyphs and were used by the early Egyptians as a way of designating numbers. Individual glyphs were combined in long sequences to make up large numbers.*

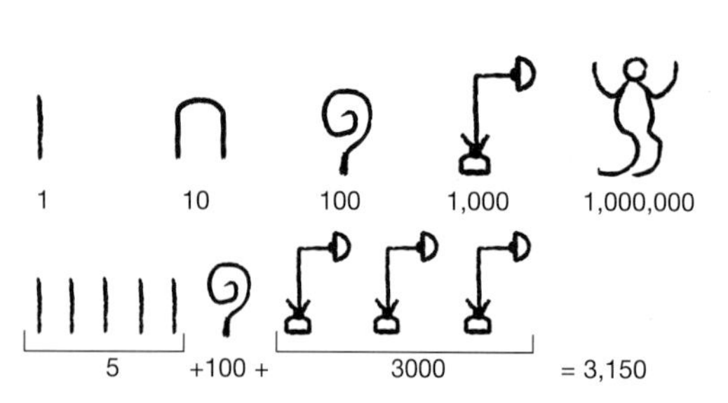

## Preparation for Life Eternal

The painting at right comes from the tomb of Inherkha in Thebes. It shows a priest wearing a mask offering a mummy a bowl of holy water. The ancient Egyptians believed that the soul continued to live after death, but that in order to survive it needed its physical body as well. As a result, the Egyptians placed great importance on the preservation of the body after death.

The word mummy comes from the Arabic word *mumiyah*, which means "bitumen." It was once believed that bitumen, a tarlike substance, was used in the embalming process to preserve corpses. In fact, we now know that the main ingredient used was natron, which is hydrated sodium carbonate. The Egyptians would have obtained it from the dry beds of salt lakes.

The process of mummification was a complex ritual. The Greek historian Herodotus, who lived in the fifth century B.C. and visited Egypt, described the process in his work *The Histories*:

"First they draw out the brains through the nostrils with an iron hook, raking part of it out in this manner, the rest by the infusion of drugs. With a sharp stone they make an incision in the side, and remove all the bowels; having cleansed the abdomen and rinsed it with palm wine, they next sprinkle it with pounded perfume. Then, having filled the belly with pure myrrh, cassia, and other perfumes, they sew it up again; and then steep it in natron, leaving it under for seventy days... [then] they wash the corpse, and wrap the whole body in bandages of waxen cloth, smearing it with gum... After this the relatives, having taken the body back again, make a wooden case in the shape of a man, and having made it they enclose the body; and then, having fastened it up, they (put it in the) sepulchral chamber, setting it upright against the wall. In this manner they prepare the bodies that are embalmed in the most expensive way."

## GREEKS, ROMANS, AND BYZANTINES

The arrival of the Greeks in Egypt marks a definite break from what we think of as "ancient Egypt." The Greeks brought new values and a powerful new culture

### Greek influence

The Egyptians welcomed Alexander the Great as a liberator, freeing them from the yoke of the Persian empire. He founded a new capital on the Mediterranean coast and named it Alexandria after himself. After his death in 323 B.C., Alexander's empire was divided up among his generals. Egypt was claimed by Ptolemy (305–282 B.C.), and Ptolemy's descendants—the Ptolemaic Dynasty—ruled the country until 30 B.C.

*This sarcophagus from the fourth century A.D. combines Egyptian and Roman styles: the former in the mummy case itself, the latter in the delicate painting of the woman's face.*

Under the Ptolemies, Alexandria grew into a great Mediterranean city. It was a center of Greek learning, with a library famed throughout the ancient world. The new Greek rulers respected Egyptian beliefs—many new temples were built and old ones restored, and Greek gods were merged with the Egyptian ones. But corruption and rivalries within the ruling classes weakened the dynasty's grip on power just as the riches of Egypt were attracting the attention of the expanding Roman empire.

### The Roman Province

The decline of the Ptolemaic Dynasty in Egypt coincided with the rise of the Roman republic and a period of conflict and conspiracy in both Egypt and Rome. Cleopatra VII, the last of the Ptolemaic Dynasty, was queen of Egypt from 52 to 30 B.C. She was an ambitious and charismatic ruler who tried to strengthen Egypt's position by forming alliances with powerful Romans. She claimed to have had a son with Julius Caesar when he visited Egypt in 48 and 47 B.C. After Julius Caesar was murdered in Rome, Cleopatra turned to a new protector.

# THE ROMAN EMPIRE

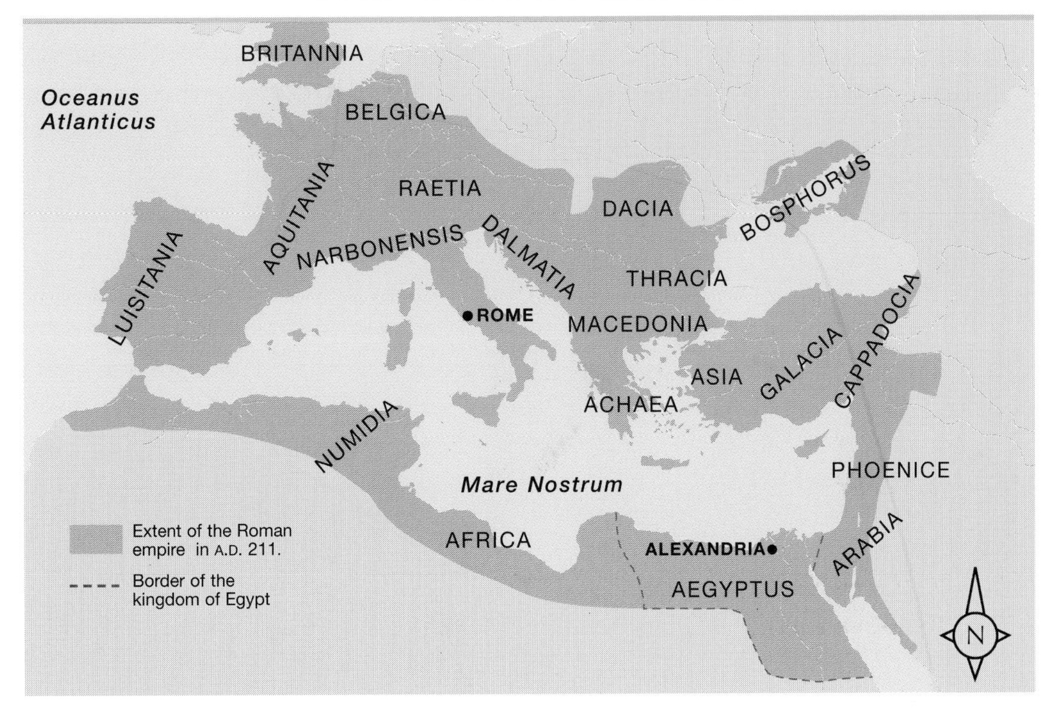


***The map above shows the Roman empire at the height of its power in A.D. 211. Egypt (Aegyptus) was one of the provinces of the empire, with its capital at the Mediterranean port of Alexandria.***

Mark Antony was one of the three Roman rulers who shared power in the troubled times that followed Caesar's murder. He ordered Cleopatra to appear before him in 41 B.C. at Tarsus (now in Turkey) to face charges that she had assisted his enemies. But instead of punishing the Egyptian queen, Antony fell in love with her, returned with her to Alexandria, and married her.

In Alexandria, Antony and Cleopatra led a life of extreme luxury, producing three children. They also formed a military alliance against Octavian, Antony's rival in Rome. In 31 B.C., however, Egyptian naval forces were defeated at the Battle of Actium. Antony and Cleopatra fled to Alexandria, where Octavian's army caught up with them ten months later. In despair, the tragic couple committed suicide. Antony took the soldier's way out and fell on his sword. Cleopatra clutched an asp—an Egyptian cobra and a symbol of royalty—to her breast and died of the snake's venomous bite.

*The early Christian monks, Saints Anthony and Pachomius, were Copts. The image above shows icons of the Coptic church.*

### The Christian Era

In the first decades of Roman rule Mary, Joseph, and the infant Jesus fled into Egypt from Bethlehem. Egypt began to adopt Christianity after the arrival of Saint Mark, who is believed to have begun preaching there in A.D. 40. By the end of the first century, Alexandria was a Christian city. The Egyptian Christian church, founded by the descendants of Saint Mark's converts, was known as the Coptic Church.

In the third century, the vast but unwieldy Roman empire was split into east and west. Constantinople (present-day Istanbul) became the new capital of the eastern Roman (Byzantine) empire in A.D. 330, with Rome remaining the capital of the western empire. With the rise of Constantinople, Alexandria's importance waned. Egypt itself, however, was still an important possession, providing vital grain supplies for the Byzantine capital. Egypt's history was about to be changed forever by events taking place in Arabia.

## THE ARAB CONQUEST

In A.D. 632, the prophet Mohammed, founder of the religion of Islam, declared a holy war on the Byzantine empire. Arab invaders rode into northern Egypt in 639, and within three years, the country had fallen. The conquerors established a new Arab capital called Fustat (meaning "camp"), a military garrison on the east bank of the Nile. The Arabs were tolerant rulers. They did not generally force their beliefs on Christian and Jewish Egyptians and were content to receive grain and taxes from them. During the following centuries, Egypt adopted both the Arabic language and the Islamic religion.

## Independence Regained

At first, Egypt was part of a great Islamic empire (the Abbasid empire) and was ruled by governors appointed by the Arab caliph (ruler) in Baghdad (today the capital of Iraq). This vast empire was important because it linked the markets and trade routes of the known world, from India in the east to Spain in the west. It created a civilization of considerable learning and wealth.

Egypt became independent under the reign of Ahmed ibn-Tulun (ruled 868–884), who defied the caliph in Baghdad and made Egypt his own domain. The rulers of Egypt bore the title of "sultan." Instead of sending Egyptian taxes to Baghdad, ibn-Tulun spent the revenue on improving the country's agriculture and constructing a magnificent mosque, which still stands in central Cairo. The mosque's courtyard was big enough to hold ibn-Tulun's cavalry and all its horses. After his death in 884, Egypt fell into chaos, and eventually the caliph was able to regain control and appoint another governor.

***The map shows the Abbasid empire at its height in about A.D. 800. The provinces of the empire, shown here in capital letters, were ruled by the caliph in Baghdad, until local rebellions broke out in the late ninth century.***

THE ABBASID EMPIRE IN A.D. 800

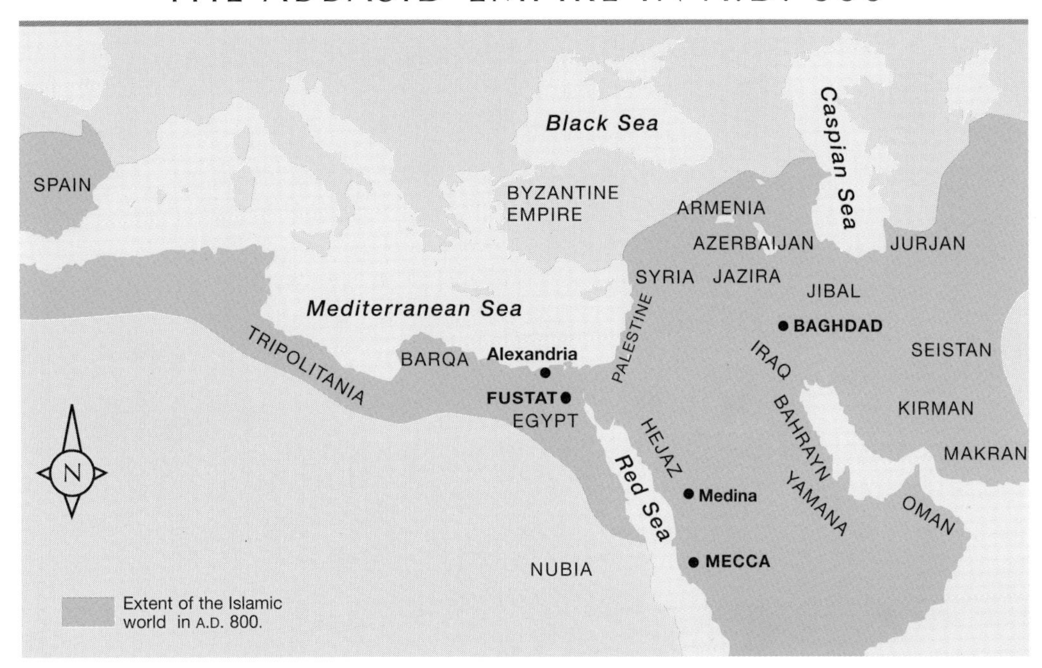

## THE FATIMID EMPIRE

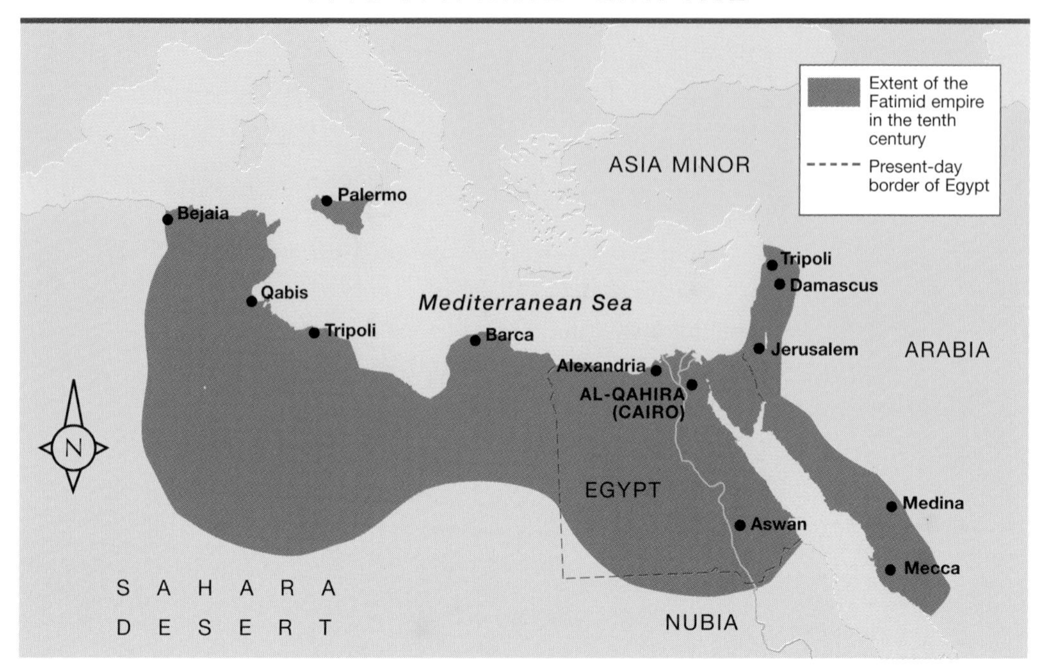


*The map above shows the extent of the Fatimid empire in the tenth century.*

**The Crusades were a series of military campaigns launched by the Pope in 1095 to liberate Jerusalem from Muslim rule. Although the Christians retook Jerusalem in 1099, the Second and Third Crusades were less successful.**

Egypt underwent a golden age during the reign of the Fatimid dynasty (969–1171). The Fatimids claimed to be the descendants of Mohammed's daughter Fatima and belonged to the Shiite branch of Islam. During the tenth century, they built themselves a mighty empire in Arab North Africa. In 969, they seized control of Egypt and set up a new capital city, al-Qahira (Cairo), just north of Fustat. They built many beautiful buildings, including the mosque and university of al-Azhar, and developed Egypt into a powerful trading country.

### Saladin and the Ayyubid Dynasty

The Fatimids were followed by the Ayyubid dynasty (1171–1250), established by the greatest of Muslim heroes, Saladin (1138–1193). In 1171, he overthrew the Fatimids and became the sole ruler of Egypt. Saladin was a champion of the Sunni branch of Islam, the branch that belongs to an orthodox tradition and believes that the first four caliphs were the successors to Mohammed. Saladin restored Egypt's relations with the rest of the Arab world. He led a holy war, or *jihad*, against the Crusaders—Christian warriors from Western Europe—who were attempting to reimpose Christian rule in areas that had once belonged to the Byzantine empire.

The Crusaders had already captured Jerusalem—a holy city for Christians, Muslims, and Jews—from the Arabs in 1099 and went on to seize most of the Holy Land (Palestine). In 1187, Saladin recaptured Jerusalem and drove the Christian invaders from all but three cities in the Holy Land. By 1291, the Crusaders had left Palestine for good.

Saladin was as brilliant a statesman as he was a military leader. He restored Egypt as a great power, fostered education, and encouraged trade and the arts. In the Islamic world, Saladin is remembered as one of the greatest heroes.

Saladin's army was largely made up of Turkish warriors called Mamluks. The word Mamluk comes from the Arabic word for "slave" and refers to those sold into slavery as children and trained to fight for the sultan. Saladin's successors became increasingly dependent on the Mamluks to keep order in Egypt and to protect the country from foreign invasion. They gained so much power that, in 1250, they were able to overthrow the Ayyubid dynasty and appoint one of their own as sultan.

Under the Mamluk sultans (1250–1517), Egypt became the center of the Islamic world. Unlike earlier dynasties, the Mamluk sultanate did not pass from father to son. Usually the throne was seized by force after the death of the sultan. In all there were 45 Mamluk sultans.

## Saladin

Born into a successful Kurdish family (from the east of modern Turkey), Saladin was brought up in Syria, where he first served under his uncle. In 1169, at the age of 31, he became commander of the Syrian troops and vizier (ruler) of Egypt. Using Egypt as his base, Saladin united Syria, Palestine, northern Mesopotamia (present-day Iraq), and Egypt. His recapture of Jerusalem in 1187 from the Crusaders was followed by the frustration of the Christian counterattack, the Third Crusade, which failed.

One of the great achievements of the Mamluks was their success in resisting the invasions of the Mongols. These were a warlike people from eastern Asia who, in the 13th century, built the largest land empire the world had ever known. It stretched from the Pacific Ocean to the eastern borders of Europe. The Mamluks were also important patrons of religion and art. Islamic architects built numerous mosques, palaces, colleges, religious schools, and hospitals. Islamic scholars compiled vast encyclopedias and dictionaries.

This 15th-century Arabic manuscript shows Mamluk warriors on horseback.

The Mamluks, though, were far less tolerant than had been the earlier Arabic dynasties. Previously, Coptic Christians had enjoyed a privileged position in Egypt and were often among the wealthiest people in society. Under the Mamluks, Muslims and Christians came into conflict. The sultans closed Christian churches. The Coptic language fell into disuse, and many Copts converted to Islam.

## OTTOMAN EGYPT

Meanwhile, the Ottoman Turks were building another empire, with its capital at Istanbul—the city that, as Constantinople, had once been the capital of the Byzantine empire. In 1516, under the command of their last sultan, al-Ghourri, the Mamluks fought a battle against the Ottomans at Aleppo, in Syria. The Mamluks were crushed, and in 1517, the Ottoman sultan Selim I rode into Cairo. Now a far-flung province of the Ottoman empire, Egypt fell into decline. The Ottomans were happy to collect their imperial taxes and leave the administration of the country to the Mamluk lords, known as *beys*. In 1796, one *bey* became powerful enough to seize control in Cairo.

# THE OTTOMAN EMPIRE

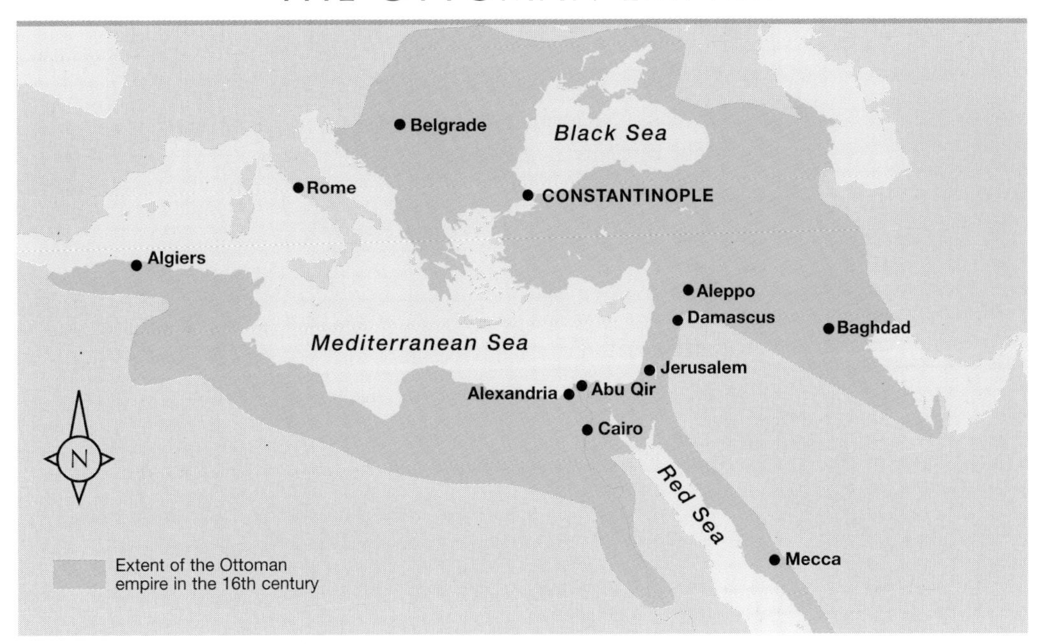


## Napoleon's Invasion

The next major event in Egyptian history was caused not by local events but by wars between France and Britain at the end of the 18th century. A French invasion force led by Napoleon Bonaparte (1769–1821) landed at Abu Qir in 1798 and swiftly captured Alexandria and Cairo. Napoleon's intention was to take control of the trade routes to the Red Sea and the Indian Ocean, disrupting British trade to the east and threatening Britain's colony, India. Napoleon brought with him a troop of scholars and artists, whose project was to study and record Egypt's monuments, arts, and plants and animals. The French occupiers discovered Egypt's magnificent monuments on the banks of the Nile, and this started a lasting European fascination with ancient Egypt (*see* p.68).

*The map above shows the Ottoman empire at its greatest extent in the 16th century. After Constantinople (modern Istanbul) was captured in 1453, the capital of the empire was moved there.*

Napoleon's supply lines were cut when British warships under Admiral Lord Nelson defeated the French fleet at the Battle of the Nile off the coast of Alexandria in 1798. The British also began to negotiate with the

## Egyptomania

The scholars and artists who came with Napoleon on his imperial expedition to Egypt studied and recorded the country's monuments, arts, and plants and animals. Their findings were published in a massive book called *Description of Egypt,* which ran to 24 volumes and was illustrated with some 3,000 drawings.

The book was the beginning of Europe's fascination with all things Egyptian and became the foundation of Egyptology. Many foreigners went to see the country's sights for themselves, including American writer Mark Twain (1835–1910) and the British nurse Florence Nightingale (1820–1910). A British tourist agent first set up voyages along the Nile in the 1860s. Novels and operas with ancient Egyptian settings, described temples, tombs, and glorious treasures. Painters produced pictures of what they claimed to be contemporary Egypt, although many had never even been to the country. The paintings showed busy markets and ornate palaces and were often concocted in artists' studios from fake Egyptian props. Buildings everywhere were decorated with sphinxes and other Egyptian motifs.

It was during this time, too, that many museums in Europe and America began to build up collections of ancient Egyptian art— like that at the Belvedere Castle in Vienna, Austria, shown above. Archaeologists and treasure-hunters dug up the treasures and sent them back to their countries, often without permission from the local Egyptian authorities. Today, the Egyptian government is demanding the return of some of the most famous artifacts, including the Rosetta Stone (in the British Museum in London, *see* p.102) and a beautiful sculpture of Nefertiti (in the Berlin Museum).

Ottoman Turks, and the Egyptian people began to rebel against their new rulers. The short-lived French occupation ended in 1801, and the Ottomans regained control.

**The French and British fought each other for control of Egypt throughout the 19th century. Egypt served as a quick route to the eastern colonies of the two powers.**

## Mohammed Ali's Modernization

Mohammed Ali (ruled 1805–1848) was a lieutenant in an Albanian contingent of the Ottoman army, which occupied Egypt in 1801. Following a long power struggle among Mamluks and Ottomans, Mohammed managed to get himself appointed pasha—or viceroy—of Egypt in 1805, ruling in the name of the Ottoman sultan in Istanbul. The Ottoman empire was weak and in decline, so in reality, Mohammed was the ruler of an independent country. He tightened his grip on power by cleverly playing off his opponents against each other, taking advantage of the people's dislike of their Turkish overlords. In 1811, he ordered the massacre of the Mamluk leaders in Cairo.

Mohammed Ali's goal was to completely modernize his country, using European countries such as France and Britain as his models. He introduced a public education system, reformed the army and administration, and set up a government printing press. His attempt to turn Egypt into an industrial nation failed because of the country's lack of energy sources.

*The image below shows Mohammed Ali, as painted by a French artist in 1811.*

Mohammed's successors continued his attempts to modernize and Westernize the Egyptian economy. Foremost among these projects was the building of the Suez Canal. In 1854, the pasha Sa'id gave the French permission to cut a canal through the Suez Isthmus (the narrow strip of land between the Sinai desert and the rest of Egypt). However, work did not begin until the reign of his successor, Isma'il (ruled 1863–1879), the grandson of Mohammed

## An Egyptian Empire

Mohammed Ali and his successors attempted to build an empire that rivaled the lands controlled by the Ottoman sultan. Ali took lands around Mecca and Medina in Arabia and northward into Syria, although these were recaptured in 1840. Under Isma'il, Egypt extended its borders south into Sudan and along the Red Sea coast as far as Somalia. This gave Egypt control of a substantial slave and ivory trade. The lands to the south of Egypt were lost with Isma'il's abdication in 1879.

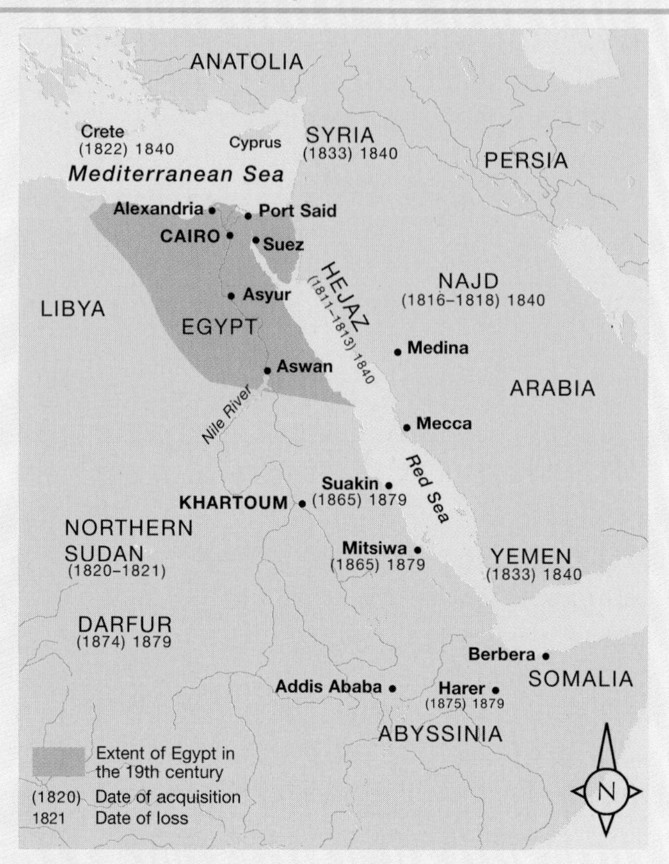


Ali. The French empress Eugénie finally opened the Suez Canal in 1869 (*see* p. 90), in the presence of several European monarchs.

Railroads, factories, bridges, and irrigation systems were also built, usually with British or French help. The Egyptian cotton industry flourished, especially during the American Civil War, when U.S. cotton production collapsed. Despite these advances, Isma'il ran up huge national debts to France and Britain. He attempted to pay off these debts by selling shares in the Suez Canal to the British. However, these attempts were unsuccessful and in 1879 the Ottoman sultan, under pressure from Britain and France, forced Isma'il to give up the throne.

## THE BRITISH OCCUPATION

The instability that followed Isma'il's forced abdication led the British to intervene in 1882 to protect their trading routes and to ensure the repayment of their loans. They quickly occupied Suez and Cairo. Although Isma'il's son Tawfiq was made the official ruler of Egypt, all the important political decisions were made by the British occupiers. The Egyptian people resented British rule deeply and wanted to set up a fully independent nation, free from any foreign interference, whether it be from Europe or Istanbul.

With the outbreak of World War I in 1914, Britain went to war with Ottoman Turkey, which entered the war on the side of Germany. Britain imposed martial (wartime) law on Egypt and made the country a British protectorate. After the war, the emergence of a popular Egyptian nationalist party—the Wafd—led the British to grant a measure of independence to Egypt on February 28, 1922.

*The image above shows Egyptians rioting against European businesses in Alexandria in 1882.*

### Toward Independence

A constitutional monarchy was created. King Fuad I (ruled 1922–1936) was made head of state, and a new government of elected and appointed members was set up. Britain, however, was not prepared to lose control of the Suez Canal and still maintained a firm grasp on Egyptian affairs. For three decades, a struggle for influence was waged between the British High Commissioner, King

*Lord Allenby, British Lord High Commissioner for Egypt, at work on the veranda of his residency in Cairo in 1924. Allenby worked on the reorganization of the new Egyptian government following political and military clashes between Britain and Egypt.*

Fuad and his successor, King Farouk (ruled 1936–1952), and the Wafd. During World War II (1939–1945), Egypt played a vital role as a base for British military operations in the Middle East and North Africa. The British victory over the Germans at the Battle of El Alamein, which took place in the desert west of Alexandria, was a major turning point in the war.

The great majority of Egyptians did not support Britain during the war, and the postwar period was one of political instability and popular unrest. Egyptian forces were defeated in the Israeli War of Independence in 1948, and the next few years were marked by terrorism, assassinations, and increasingly violent anti-British demonstrations.

## INDEPENDENCE

In July 1952, a group fighting for national independence called the Free Officers, led by Colonel Gamal Abdel Nasser (1918–1970), overthrew King Farouk in a bloodless coup. Born in a poor quarter of Alexandria, Nasser attended military college and served in the Egyptian army in the war with Israel in 1948. Nasser was very popular with the great mass of Egyptians. After centuries of foreign domination, he was the first true Egyptian to rule the country since the pharaohs of the fourth century B.C.

## The Suez Crisis

Nasser became prime minister and then president of the new Arab Republic of Egypt. Under his leadership, Egypt distanced itself from any alliance with either the Western powers, such as Britain and the United States, or with the Soviet Union. In 1956, Britain and the United States showed their discontent with the turn of events in Egypt by withdrawing their offer of funding for the High Dam project at Aswan. Nasser retaliated by nationalizing the Suez Canal Company to raise the money for the Aswan project. This meant that the company was taken back under the control of the government and all revenue from ships passing through the canal went to the state rather than to the previous owners of the canal, which was part owned by the British and French governments. Nasser aimed to use the Aswan High Dam to improve Egypt's agriculture and replace its colonial cotton industry with a more diverse, modern economy.

The trade of both Britain and France depended on their free access to the canal. Nasser's action led Britain and France to cooperate with Israel in mounting a military attack on Sinai and Suez, with the intention of taking control of the canal zone. Although the Egyptian forces were defeated,

***The British minelayer* HMS Manxman *passes through the Suez Canal after its seizure by British and French troops in 1956. The Egyptians sank some ships in the canal to block access.***

## THE MODERN MIDDLE EAST

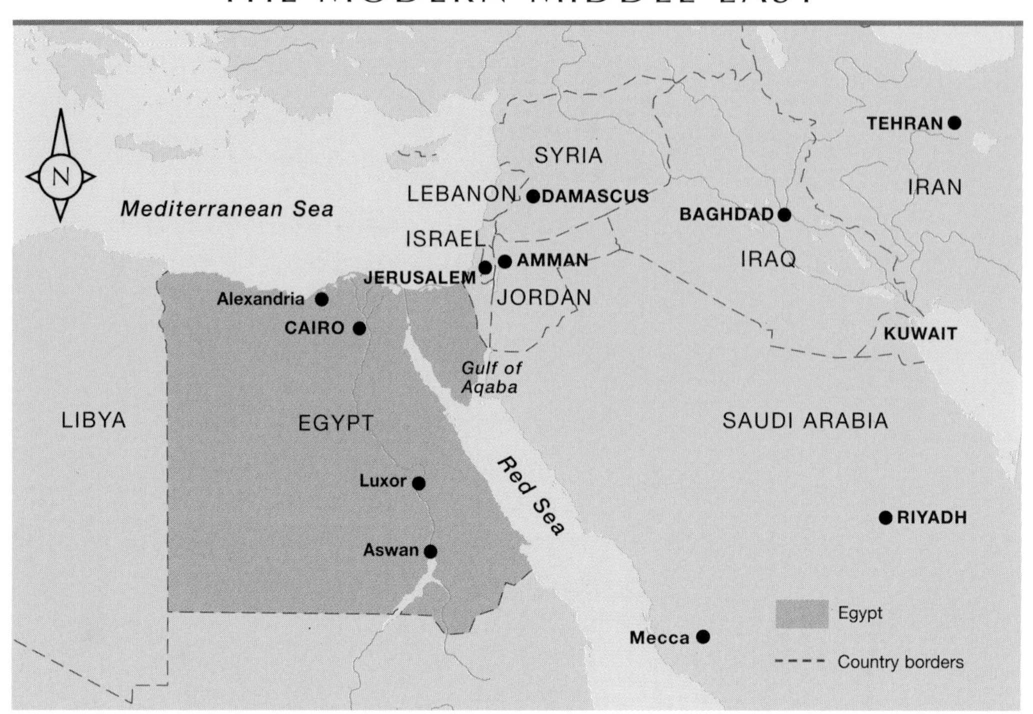


***Today Egypt is the geographical link between the Arab states of the Middle East and those of North Africa. The country's borders have remained unchanged since the return of Sinai from Israel in 1979.***

both the United States and the Soviet Union opposed the action and put pressure on Britain and France to withdraw troops. Despite military defeat, Egypt won control of the canal and claimed a triumph over its enemies.

President Nasser's defiance of the West made him a hero to Arabs throughout the Middle East. He dreamed of becoming leader of an alliance that would embrace the entire Arab world. To this end, he succeeded in creating a short-lived United Arab Republic that included Egypt, Syria, and, briefly, Yemen. The union dissolved in 1961 and his dream was never realized.

### The Six-Day War

Meanwhile, another crisis was brewing. The creation of Israel by the United Nations in 1948 as a homeland for the Jewish people caused resentment in the Arab world, but especially with Israel's closest Arab neighbors, Jordan and Syria. Egypt, along with other Arab nations, sent soldiers to help the Palestinians fight the new Jewish state. Arab-Israeli tension peaked in early 1967, when Egypt and Jordan pledged support for Syria in any conflict with Israel. Nasser closed the entrance to the Gulf of Aqaba, cutting off Israeli shipping from access to the Red Sea.

The Israeli government saw Nasser's closing of the strait as a kind of declaration of war. Believing themselves to be surrounded by hostile forces, they launched a military assault on Egypt on June 5, 1967. Within six days, the Israeli military destroyed most of the Egyptian air force on the ground, captured Sinai and the Suez Canal, took the West Bank territories from Jordan, and occupied the Golan Heights in Syria.

**The hostility of Israel's Arab neighbors led Egypt to build up strong military forces. Israel was supported by the United States, while Arab states, including Egypt, were supported by the Soviet Union.**

By the time the United Nations managed to impose a cease-fire on June 11, almost 10,000 Egyptians were dead and the towns along the Suez Canal lay in ruins. The country had suffered a crushing defeat.

President Nasser resigned, but the Egyptian people showed their refusal to accept his resignation with mass demonstrations. Nasser remained as president until his death of a heart attack in 1970. Despite military defeats

*Israeli tanks drive across the recaptured Sinai Desert on the second day of the Six-Day War in 1967. The Sinai was only returned to Egypt in 1979.*

and a worsening economy, Nasser had given Egypt back to the Egyptians and ended the indignity of foreign rule. He was and remains a hero in Egypt, and many homes and shops still proudly display his photograph on their walls. The Suez Canal remained closed until 1975, and the territory lost in Sinai was not recovered until after a peace treaty between Egypt and Israel signed in 1979 by Nasser's successor, Anwar el-Sadat.

***Born in 1918, Anwar el-Sadat was involved in Nasser's campaign to oust British rule from Egypt. On succeeding Nasser as president, he expelled Soviet advisers and assumed military control. His dramatic journey to Israel in 1977 and the resulting peace treaty earned him the Nobel Peace Prize.***

## Egypt and Israel

Sadat (ruled 1970–1981) had been Nasser's vice-president. He introduced Western ideas to Egyptian society and sought to repair its damaged economy by improving relations with the United States and Western Europe. He surprised the world, however, by attacking Israel in the October War—also called the Yom Kippur War—of 1973. Although Egypt and its Arab allies were rapidly defeated by Israel's U.S.-backed military superiority, Sadat used the war as a bargaining point to sue for a long-term peace with Israel.

In March 1979, Sadat signed a dramatic peace treaty with Israel's prime minister, Menachem Begin. Israel withdrew from Sinai in return for Egypt's recognition of Israel's right to exist, and a United Nations (U.N.) peacekeeping force was installed to police the Egypt–Israel border. The agreement also resulted in Egypt receiving billions of dollars in foreign aid from the United States.

Peace with Israel brought Egypt alienation from the rest of the Arab world. Egypt was expelled from the Arab League and was not readmitted until 1989.

Despite domestic approval of Sadat's liberalizing policies and the peace with Israel, there was growing dissent at the unequal distribution of wealth in the new economy and at the plight of the stateless Palestinians. During a military parade in Cairo on October 6, 1981, a member of the terrorist group *el-Jihad* (*see* box below) assassinated Sadat.

## Islamic Fundamentalism

Islamic fundamentalism is a political and religious movement that is an important force throughout the modern Arab world. It combines traditional Muslim values based on *Shari'ah*—the law of Islam—with revolutionary action. Islamic fundamentalism is often hostile to Western-style secular (nonreligious) lifestyles and to the political ideas and systems that have been introduced from the West since colonial times.

In the 1980s and 1990s, many Muslim countries made concessions to Islamic fundamentalists in areas such as education, law, dress, and behavior. For example, many Arab countries have introduced laws requiring women to cover their heads and faces with a veil.

The first Islamic fundamentalist group, the Muslim Brotherhood, was founded in Egypt in 1928 by Hasan al-Banna. During World War II, membership grew to two million and it began to pose a threat to the Egyptian political system. Branches were established in other Muslim countries. In 1954, however, the new Free Officer regime in Egypt suppressed the Brotherhood, and its membership declined until the 1970s.

The fundamentalist movement's revival came during and after the defeat suffered by Egypt in the Six-Day War with Israel in 1967. It also benefited from wealthy Arab countries, such as Saudi Arabia, which supported Islamic causes. It drew much of its support from the growing numbers of Arabs who lived in large cities. Powerful fundamentalist movements developed in many Muslim states such as Iran, Algeria, Pakistan, Afghanistan, and elsewhere. Under President Sadat, the Muslim Brotherhood re-emerged and a number of terrorist attacks on government buildings preceded Sadat's assassination in 1981.

In recent years, President Mubarak's regime has clamped down on Islamic fundamentalist groups such as the *Gama'a al-Islamiya* (Islamic Group) and *el-Jihad* (Holy War), which have been involved in violent clashes with Coptic Christians and have launched attacks on foreign tourists. Religious parties have also been banned from the People's Assembly.

### The 1980s and 1990s

The country's vice-president, Hosni Mubarak (born 1928), succeeded Sadat. Mubarak was faced with the problems of an economy staggering under the weight of an exploding population. He also had to face growing support for Islamic fundamentalism (*see* p. 77). The Middle East peace process was floundering, and Egypt's relations with Israel were severely tested after the Israeli invasion of Lebanon in 1982.

**The Arab League of 21 nations was set up in 1945 to foster the political aims of the Arab world. When Egypt was expelled from the league for recognizing the State of Israel, the headquarters moved from Cairo to Tunis.**

President Mubarak continued Sadat's policy of cooperation with the West, but he also managed to repair relations with the Soviet Union and Syria. The Arab League's headquarters was moved back to Cairo from Tunis in 1990. In the Gulf War of 1990, when Iraq invaded its neighbor Kuwait and threatened Saudi Arabia, Mubarak brought Egypt into the U.N. alliance that struck back at Iraq after its invasion of Kuwait.

Trouble flared on Egypt's southern border in the early 1990s when conflict arose with neighboring Sudan over a disputed region in the remote southeastern corner. Mubarak survived an assassination attempt in 1995, which, Egypt claimed, was sponsored by the Muslim fundamentalist Sudanese government.

## EGYPT'S GOVERNMENT TODAY

Egypt's 1971 constitution describes the country as "an Arab republic with a democratic, socialist system." There is one legislative (law-making) house, the People's Assembly, with 454 members. Of these, 444 are directly elected by the people and 10 are appointed by the president. There is also a 210-member Advisory Council, of whom 140 are elected by the people and 70 are appointed by the president.

The head of state is the president, who is nominated by the People's Assembly. This nomination is then confirmed or rejected by a popular referendum. The president serves for terms of six years. The head of government is the prime minister, who is appointed by the president,

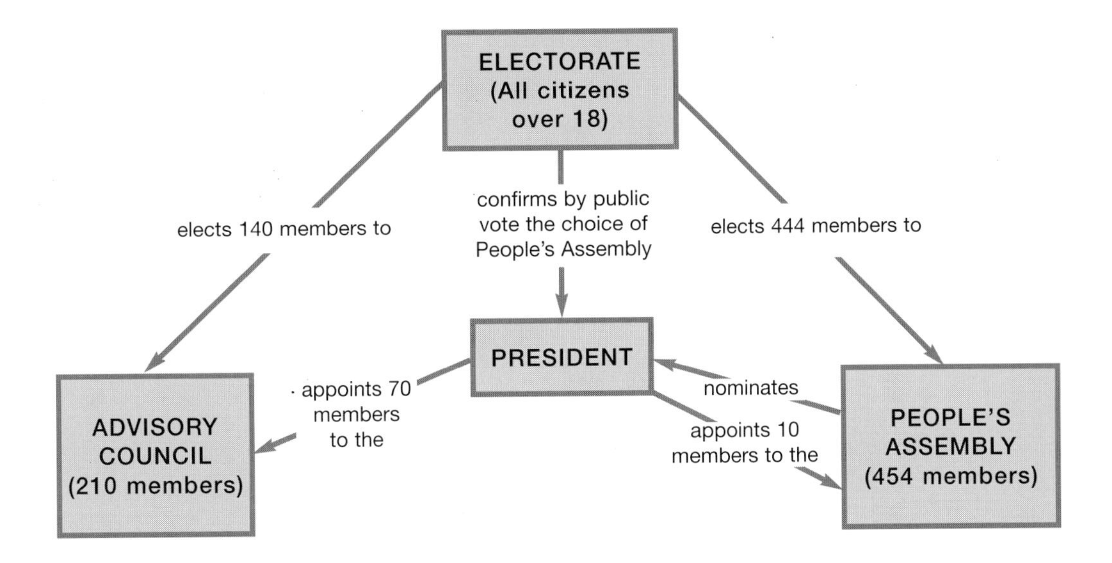


and is usually the leader of the party with the most seats in the People's Assembly. The current prime minister is Kemal Ahmed al-Ganzuri, who was elected in 1996.

***While Egypt is a democracy, the Egyptian president holds huge political power.***

## Political Parties

The National Democratic Party (NDP) enjoys a massive majority in the People's Assembly; smaller parties include the Wafd and the Progressive Rally. The government is engaged in resisting the rise of Islamic fundamentalism. The Muslim Brotherhood has gained support in the country by providing education and health systems based around the mosques. Although the Brotherhood is the most effective opposition, like other religious parties, it is technically illegal and banned from the People's Assembly.

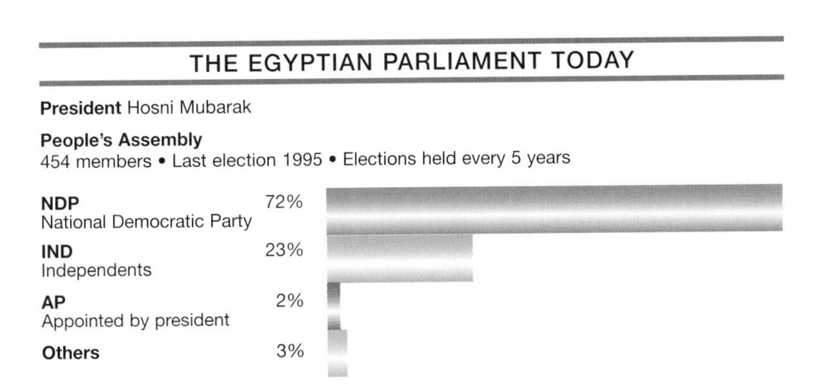


***The ruling NDP enjoys a massive majority and has close links with the military. The current state of emergency bans religious parties from sitting in the People's Assembly.***

# The Economy

*"An Arab republic with a democratic, socialist system."*

Description of Egypt in the country's constitution

Following independence in 1952, President Nasser began the process of moving Egypt toward a planned economy. The Suez Canal was nationalized in 1956, and during the 1960s and 1970s, most of industry, agriculture, transportation, and energy production was taken into public, or state, ownership. Although this brought many improvements for Egypt's poor, it also meant that foreign investors deserted the country as their assets were seized by the Egyptian state.

By the late 1980s, Egypt was suffering from poor economic management, high inflation, and low productivity, despite financial aid from oil-rich Arab nations in the 1970s and, after the 1979 peace treaty with Israel, from the United States. These problems were made worse by the high rate of population growth—more than a million a year—terrible overcrowding in the cities, and high levels of unemployment.

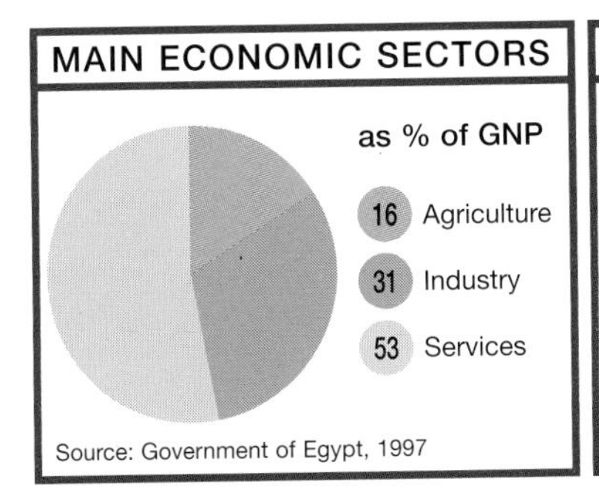

***The Aswan High Dam in Upper Egypt provides roughly 20 percent of the country's energy and the vast majority of its hydroelectric power.***

## FACT FILE

- Egyptian agriculture uses around 1,900 billion cubic feet (53.1 billion cu m) of water each year—more than 95 percent of the water supplied by the Nile River.
- The transit fees on the Suez Canal provide Egypt's largest single source of foreign currency—more than $2 billion a year.
- The Egyptian government aims to increase the proportion of habitable land in Egypt from 6 percent in 1997 to 20 percent by 2017. This is to be done by settlements in Sinai and at the oases of the Western Desert.

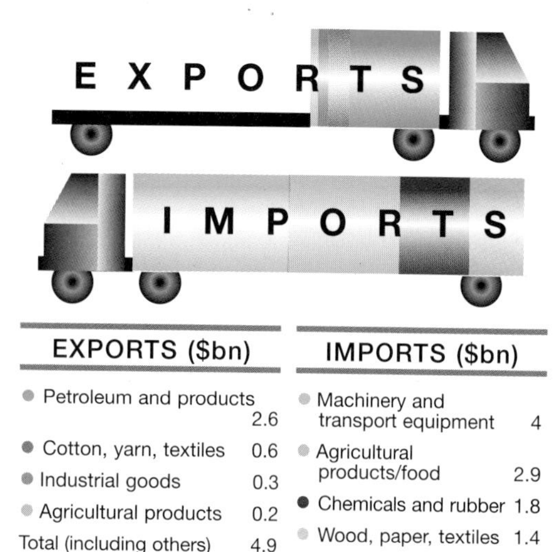


| EXPORTS ($bn) | |
|---|---|
| Petroleum and products | 2.6 |
| Cotton, yarn, textiles | 0.6 |
| Industrial goods | 0.3 |
| Agricultural products | 0.2 |
| Total (including others) | 4.9 |

| IMPORTS ($bn) | |
|---|---|
| Machinery and transport equipment | 4 |
| Agricultural products/food | 2.9 |
| Chemicals and rubber | 1.8 |
| Wood, paper, textiles | 1.4 |
| Total (including others) | 14.7 |

Source: Government of Egypt, 1997

In response to these pressures, Egypt is now moving away from its old socialist policies and toward a free-market economy, where businesses are privately owned rather than being owned by the state. A privatization program that began in 1993 has seen many government-owned companies sold off into private hands, particularly in the industrial and manufacturing sector. The formation of new, small businesses is also encouraged. Inflation (the rate at which prices increase) has dropped from about 30 percent in the 1980s to less than 10 percent in the late 1990s.

***Egypt imports a lot more than it exports. The economy is supported by large-scale foreign aid.***

Economic reforms and development have been helped by the continuing peace with Israel, with whom the country now does a significant amount of its trade. Lower spending on defense means that more money can be invested in industry and public services such as transportation. Foreign aid and investment continue, especially from the United States. Egypt's political and military contribution to the 1990 Gulf War was rewarded by one-quarter of its national debt being written off. The country also received a huge injection of aid from Saudi Arabia for the support it gave.

Source: Government of Egypt

Egypt's principal exports are petroleum, cotton (raw cotton, cotton yarn, textiles, and clothing), aluminum, steel, and citrus fruits. Main imports are wheat, maize, dairy products, chemicals, iron, machinery, and transportation equipment.

## MAIN ECONOMIC SECTORS

Egypt's main economic revenues come from oil and gas, although the country also exports substantial amounts of agricultural produce. Other important sectors are tourism and light manufacturing industry; the country also generates income from tolls on the Suez Canal and on earnings from Egyptians working outside the country, largely in the Gulf States, such as Saudi Arabia. Under President Mubarak, Egypt has been attempting to increase manufacturing in order to support its rapidly expanding population.

**Egypt's national debt is just under $30 billion. Paying interest on this debt creates the need for hard foreign currency. As a result, much of the economy is geared toward the export market.**

### Farming

Agriculture plays a major part in the Egyptian economy. In 1995, it accounted for 22 percent of gross domestic product—the total value of a country's annual output of goods and services. It employed around 30 percent of the workforce and accounted for around 20 percent of export earnings. Since Egypt's farming acreage is so small, irrigation is extremely important. More than one crop has to be grown on the same land each year, and

*The flooding of the Nubian Valley to form Lake Nasser has led to the depletion of fertile silts in the Nile Valley and to a reliance on imported chemical fertilizers.*

## HOW EGYPT USES ITS LAND

*The vast majority of Egypt is desert. Pasture is restricted to the coastal area in the northwest.*

formerly unproductive land must be brought into use. Almost 20 percent of all farmland is now reclaimed desert.

Egyptian agriculture is very different from that in the United States. Egypt's small area of extremely fertile, irrigated land is split up into small units and farmed very intensively, producing several crops a year. It also differs from agriculture in many developing countries, in that Egypt concentrates on commercial rather than subsistence farming (producing food to eat), growing crops that it can sell overseas for export income.

*The Egyptian government is making repeated attempts to increase the small amount of arable land in the country.*

### The Aswan High Dam

Egypt has made a huge investment in irrigation over thousands of years, but none greater than the building of the Aswan High Dam. The dam, which dwarfs the smaller Aswan Dam built by the British 4 miles (6 km) downstream, provides complete control over the waters of the Nile, on which Egypt has always depended. Farmers are no longer at the mercy of the river, which in dry years could cause the crops to fail through lack of water, and in wet years could flood catastrophically, causing damage to crops and property.

But the dam has its disadvantages, too. Now that there is no longer an annual flood, the fresh layer of fertile, black mud that was once laid down on the fields each year is no more. This means farmers must use larger amounts of artificial fertilizers—about 1 million tons each year—to keep their land productive. The creation of the dam also displaced many people in Upper Egypt.

LAND USE

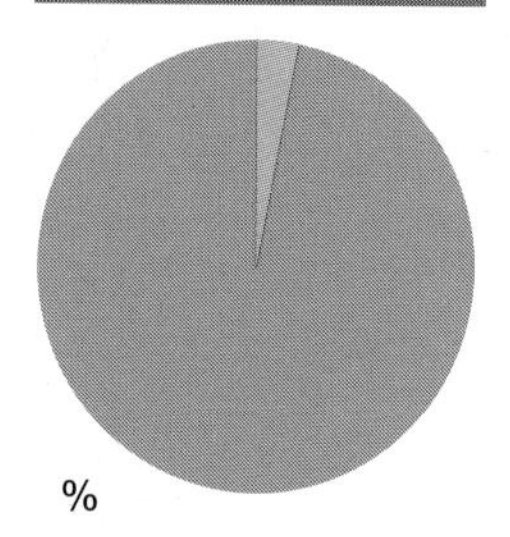

%

2 Arable land

98 Other uses

Source: Government of Egypt

Egypt's main commercial crop is cotton—it exports one and a half times as much cotton as the United States. The crop was introduced to the country in the first half of the 19th century, and Egypt is now one of the world's leading cotton exporters. Rice, onions, garlic, and citrus fruit are also grown for export. Other important crops include sugarcane, wheat, millet, beans, and lentils. Cattle, water buffalo, goats, sheep, camels, and chickens are raised mainly for local use.

**Cotton is Egypt's main summer field crop and is grown on one-fifth of available arable land. Egypt grows about 3 percent of the world's cotton.**

## Fishing

Fish has been an important part of the Egyptian diet since ancient times. Elaborate paintings on the walls of ancient Egyptian tombs often show the pharaoh fishing in the Nile using a papyrus-reed raft, watched by his wife and family.

The traditional Nile fishery has been greatly expanded by the Aswan High Dam. Fish farms have been built on Lake Nasser, and the lake has been stocked with several species of food fish. The Mediterranean shores of Egypt and the gulfs of Suez and Aqaba also support active fishing fleets. More than two-thirds of the total catch is freshwater fish.

*Stocks of saltwater fish off the Mediterranean coast, such as at Alexandria below, have become depleted by the change in the flow of the Nile River. This is largely due to the building of the Aswan High Dam in the 1960s.*

*Most of Egypt's oil production comes from fields in the Gulf of Suez. However, the bulk of the country's energy is imported.*

ENERGY SOURCES

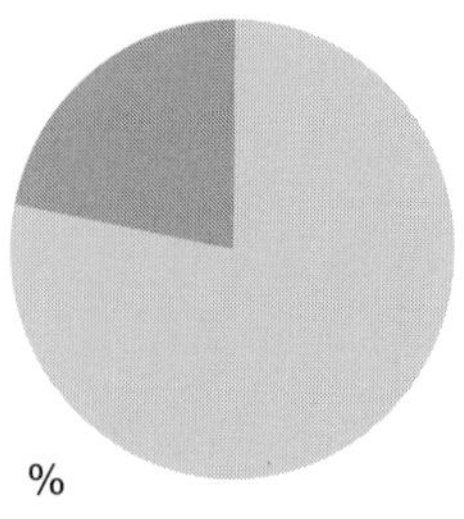

%

78 Oil, gas, coal, and diesel

22 Hydroelectricity

Source: Government of Egypt

## Mineral Resources and Energy

Most of Egypt may be barren desert, but the country has valuable reserves of mineral resources, including oil, gas, limestone, phosphates, iron ore, and manganese, and smaller deposits of chromium, uranium, coal, and gold.

Petroleum was first discovered in Egypt in 1909. However, significant oil production did not begin until the 1970s, when a number of oil fields in the Gulf of Suez and the Western Desert started production. Egypt is now a net exporter of oil (it exports more oil than it imports), though 55 percent of production is consumed domestically. Exploration is continuing in the Western Desert, the Gulf of Suez, and Sinai. Large deposits of natural gas have also been discovered.

There are outcrops of limestone over large areas of Egypt, and it is extensively quarried for use in the production of cement. Phosphate deposits from desert salt lakes are used in the manufacture of fertilizer.

The hydroelectric power-generating plant at the Aswan High Dam doubled Egypt's electricity production, but since the 1970s increasing use has been made of natural gas. Around 75 percent of Egypt's electricity was generated in gas-fired plants. Not surprisingly in one of the sunniest countries in the world, increasing emphasis is being laid on the use of solar energy.

## Industry and Services

The development of Egyptian industry in the 1960s and 1970s included the construction of iron and steel works, engineering plants, and an aluminum smelter at Aswan to take advantage of the cheap hydroelectricity produced by the High Dam. More recently, manufacturing has expanded, with factories now producing automobiles, trucks, refrigerators, washing machines, televisions, and other consumer goods. Food processing,

## MAJOR INDUSTRIES

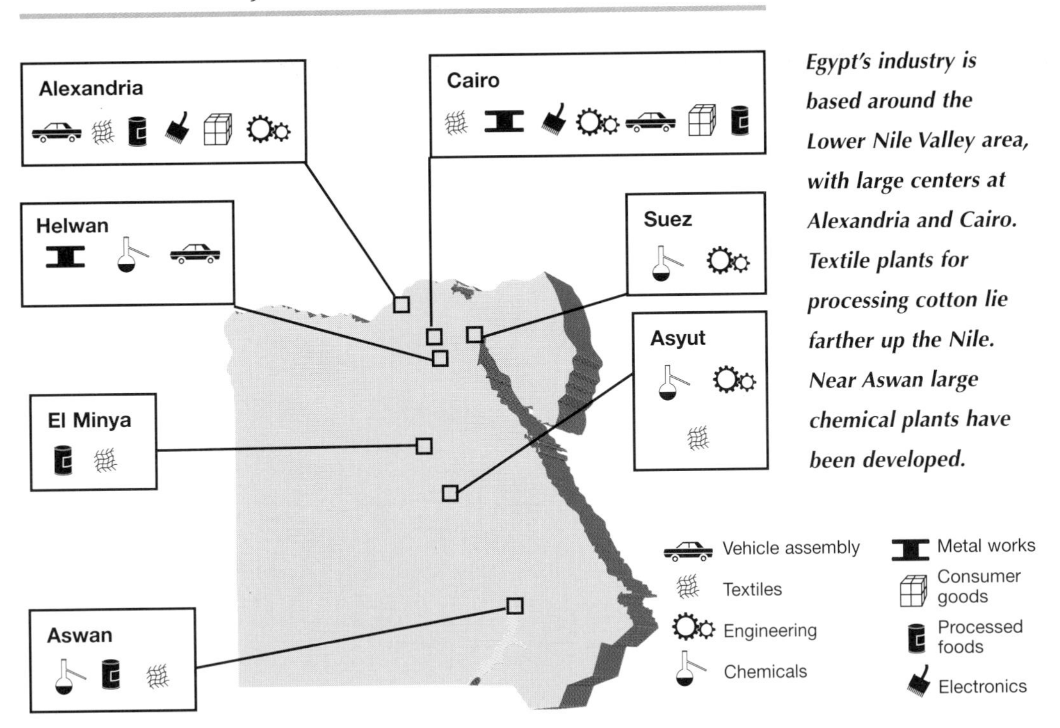


*Egypt's industry is based around the Lower Nile Valley area, with large centers at Alexandria and Cairo. Textile plants for processing cotton lie farther up the Nile. Near Aswan large chemical plants have been developed.*

*Above, a market trader in Cairo displays his bright array of fabrics draped over a car.*

oil refining, and the processing of chemicals are also becoming increasingly important.

The cotton industry was introduced to Egypt in the 19th century, and it is still an important part of the economy. In addition to selling raw cotton overseas, Egypt has processing plants and factories producing cotton yarn, cotton textiles, and finished cotton clothing. As yet, most large-scale industry is still publicly owned, with private enterprise confined mainly to textiles and food processing.

Tourism is a very important and rapidly expanding sector of the Egyptian economy. However, as with many new industries, it employs a small number of people—only 1 percent of the workforce. A record number of tourists (4.2 million) visited Egypt in 1997, earning some $3 billion for the country. Since 1997, though, the industry has been badly damaged by terrorist attacks. Fifty-eight tourists died in a particularly terrible attack by Islamic extremists at Luxor on November 17, 1997.

## Transportation

**Cairo International Airport handles 7.1 million passengers a year and the national airline, EgyptAir, is a major carrier in the Middle East.**

Historically, the Nile has been Egypt's main transportation artery for thousands of years, and it remains important today. The river itself is navigable as far as Aswan, and it is linked to Alexandria via the Mahmudiya Canal. There are many other smaller waterways in the Nile Delta. In total, Egypt has about 2,000 miles (3,200 km) of navigable waterways, about one-third of which is accounted for by the Nile. A steamer service on Lake Nasser normally links the

Aswan High Dam with Wadi Halfa in Sudan, but the service has been haphazard since relations between Egypt and Sudan deteriorated in the 1990s. The Nile plays an important part in the tourist industry.

The Suez Canal links the Mediterranean and Red seas and is one of the world's most important waterways (*see* p. 90). The transit fees on the canal are also Egypt's largest single source of foreign currency—more than $2 billion a year. The country's main shipping ports are Alexandria, Port Said, and Suez. The first railroad in Egypt was the line from Alexandria to Cairo and Suez, built by the British in the 1850s. Since then, the total length of track has increased to 3,000 miles (4,800 km). The rail network links Cairo with all the main towns and cities in the Nile Delta and the Suez

## TRANSPORTATION

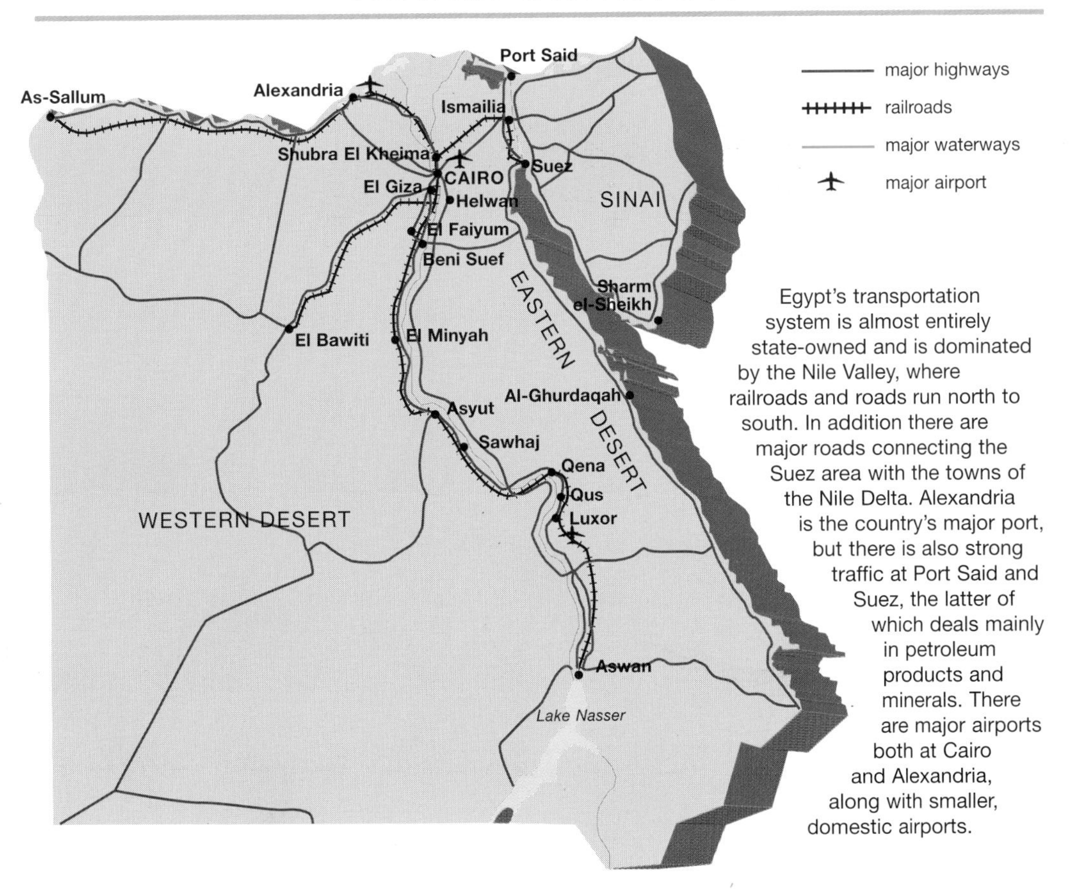


Egypt's transportation system is almost entirely state-owned and is dominated by the Nile Valley, where railroads and roads run north to south. In addition there are major roads connecting the Suez area with the towns of the Nile Delta. Alexandria is the country's major port, but there is also strong traffic at Port Said and Suez, the latter of which deals mainly in petroleum products and minerals. There are major airports both at Cairo and Alexandria, along with smaller, domestic airports.

## The Suez Canal

The Suez Canal is one of the busiest waterways in the world, providing the shortest sea route between Europe and the Indian Ocean. It stretches for 105 miles (168 km) from Port Said on the Mediterranean Sea to Suez on the Red Sea, linking Lake Manzela, Lake Timsah, and the Bitter Lakes.

In the times of the pharaohs, an irrigation channel large enough for boats to sail along was cut from the Nile toward Lake Timsah. This was later extended to the Bitter Lakes and the Red Sea, and this ancient canal survived until it was filled in for military reasons by an Arab caliph in the seventh century A.D.

The present canal was built by a French company and opened in 1869, when the channel was only 190 feet (60 m) wide at the surface, narrowing to 72 feet (22 m) wide at the bottom, with a depth of 26 feet (8 m). Since then it has been enlarged several times. It is presently being widened to 1,360 feet (415 m) and deepened to 56 feet (17 m), so that it can handle the biggest oil tankers. The canal was nationalized in 1956.

When the Suez Canal first opened, it shortened the sea route from London to Bombay, from 10,800 miles (17,280 km) (following a route around the Cape of Good Hope) to 6,300 miles (10,080 km) (via the canal). It originally carried passengers and cargo between Europe and India and the Far East. By the 1950s most of the north-bound traffic consisted of tankers carrying oil from the Persian Gulf to European refineries. Today around 50 ships a day pass through the canal, taking 15 hours to complete the passage.

Canal zone. A main line runs south from Cairo along the Nile Valley to Aswan. Another runs from Alexandria west along the coast to As-Sallum, near the Libyan border.

Egypt has about 36,250 miles (58,000 km) of highways, one-fifth of which are unpaved. The country has no superhighways. Automobile ownership in Egypt is low—only 19 cars per 1,000 people, compared to the United States, which has about 560 cars per 1,000 people. Remarkably, the country ranks eighth in the world for the number of accidents per mile traveled.

*Almost all of Egypt's major towns and cities are served by the railroads, although the system is badly in need of modernization and most routes are inferior to the more modern bus network.*

Cairo has a major international airport at Heliopolis, northeast of the city, which handles around seven million passengers a year. There are smaller international airports at Alexandria and Luxor. Charter airlines serving the tourist trade fly direct from Europe to Luxor, Aswan, Al-Ghurdaqah, and Sharm el-Sheikh. Domestic flights link Cairo to most of the country's main cities. The national airline is EgyptAir.

## Communications

Most of Egypt's telephone system is still publicly owned, although recently private companies have started to operate. It lags behind the West in terms of sophistication and reliability but is undergoing rapid development. Telephone ownership is low at around 56 per 1,000 people, compared to around 644 per 1,000 people in the United States. The postal service is fairly unreliable, with over 15 percent of letters getting lost in transit, though the service is more reliable from large cities. Television and radio broadcasting is carefully

**MAIN FOREIGN ARRIVALS**

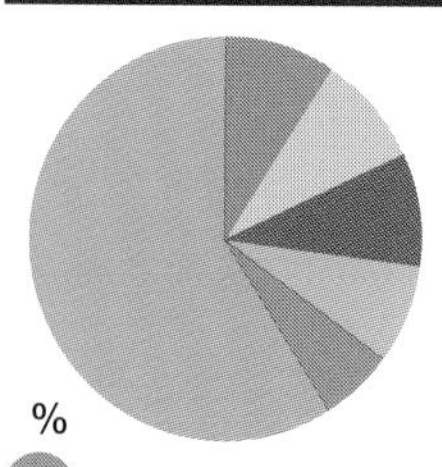

%

- 9 Saudi Arabia
- 9 Germany
- 9 United Kingdom
- 8 Israel
- 6 Libya
- 59 Others

Source: Government of Egypt

***Egypt is a popular destination for visitors from the country's Arab neighbors as well as for Europeans.***

controlled by the government (*see* p.114). The number of sets for each member of the population is low. Many people are more likely to watch television communally in a bar or social group than they are in their own homes.

## Tourism

The discovery of Egyptian treasures in the 18th century brought the first European tourists to Egypt. Many of these early visitors saw the country from the Nile, and the boats that traveled down the river were adapted to their needs with beds and cabins.

The tourism industry was given a big boost by the opening of the Suez Canal in 1869. British tour agencies took the whole tour business one step further. Instead of organizing their own accommodations, cruises, and guides, tourists would now have the whole trip planned for them in advance. The agencies' employees made efforts to ensure that the temples were open and that the tourists were provided with their own native food.

Today the Nile and its monuments have lost none of their attraction and cruise ships continue to ply the river. The traditional sights have been supplemented by more modern pleasures such as scuba diving in the Red Sea.

## The Sinai Development Project

The Egyptian government has begun a long-term project to develop the Sinai region in order to relieve the Nile Valley of overcrowding. The project will involve the creation of irrigation schemes and desalination plants, the building of new roads and towns, and the development of tourist facilities. The objective is to have some 3.2 million Egyptians living in Sinai by the year 2017. The population in 1995 was around 255,000.

A major part of the project is the construction of the Al Salam (Peace) Canal. Also known as the "Suez Syphon," this aqueduct will carry more than 3.5 billion cubic yards (32 billion cu m) of water from the Nile and irrigate more than 600,000 acres (240,000 h) of land reclaimed from the desert. It is planned that this project and associated developments will eventually increase the fertile area of Egypt from 6 to 20 percent.

In recent years the Egyptian government has become aware of the impact that its vast tourist industry has on the country's ecology and ancient monuments. Tourists are now forbidden to climb the sides of the pyramids at Giza, for example. Along the Red Sea coast the rate of hotel development has been so rapid that the government has sent in its own officials to monitor building progress.

***Over 3.7 million people visit Egypt each year, making tourism vital to the country's economy. Recently the government has tried to develop tourism in areas such as Sinai and its Red Sea resorts in order to support the local economy there.***

## Egypt's workforce

An increasingly important factor in Egypt's economy is its large workforce. Before the Gulf War in 1990, when Iraq invaded Kuwait and threatened Saudi Arabia, more than a million Egyptians worked in the Gulf States. Egypt produces the largest number of skilled graduates of any Arab country and trained Egyptian workers were much in demand across the Middle East. These workers overseas provided a boost to the Egyptian economy by sending money home to their families, but their contribution was cut short by the Gulf War. In the early 1990s, many returned home to swell the numbers of the unemployed. The drop in hard foreign currency coming into the economy had a negative effect on the country's balance of payments, and Saudi Arabia gave Egypt large amounts of aid in thanks for Egypt's support during the conflict.

It is currently unclear whether Egypt's rapidly expanding population will prove to be an asset or a burden. On the one hand the increasing number of people puts pressure on the already overstretched resources of the country. On the other hand it is a stimulus for the government to expand the economy, and a population roughly the size of Britain's or France's will wield increasing power in the world economy, encouraging investment from overseas.

شفامی دهد که معاودت صورت نبندد ومن جمله این مقدمات از علم طب

# Arts and Living

*"Literature should be more revolutionary than revolutions themselves."*

20th-century Egyptian writer Naguib Mahfouz

Despite the 3,000-year reign of the pharaohs, three centuries of Greek rule, and seven centuries of Byzantine domination, modern Egyptian culture is predominantly Arabic and Islamic. Egyptians are understandably proud of their country's rich and varied cultural heritage. The ancient Egyptians painted many paintings and sculpted impressive statues. The traditions of storytelling and music that began during ancient times are still highly valued art forms in modern Egypt. The country's 20th-century struggles for independence helped this artistic tradition to develop for the modern age.

Today, Egypt dominates the arts in the Arab world. The government plays an important role in ensuring that many branches of the arts continue to thrive. Government-run organizations help make films, stage theatrical productions, and set up art exhibitions. The state (government) is also responsible for the artistic centers that teach and show cultural disciplines. In Cairo, the Academy of Art teaches young dancers, filmmakers, actors, and musicians the skills they need in order to pursue careers in the arts.

Egypt is the center for publishing and filmmaking in the Arab world, and as a result, Egyptian culture has spread throughout the Middle East. It is also slowly becoming more appreciated in the West.

***The Egyptian tradition of miniature painting was influenced by the country's links with Persia and India. Here a doctor is shown tending to a patient.***

FACT FILE

- In 1988, Naguib Mahfouz became the first Arabic-language writer to win the Nobel Prize for Literature. Today, only about half of his novels and short stories have been translated into English.
- During the 1940s and 1950s, Cairo's film studios produced around 100 films a year. Today, only about 20 films are made in Egypt each year.
- On the death of Egypt's most popular singer, Umm Kolthum in 1975, the city of Cairo was brought to a standstill as millions of grieving Egyptians poured onto the streets. Her recordings are still heard all over Egypt.

## Ancient Egyptian Art and Architecture

The first monumental (large-scale) buildings were made in Egypt some 5,000 years ago, when King Zoser began the construction of his famous Step Pyramid at Saqqara. Over the next few thousand years, ancient Egypt's skilled and daring architects created some of the most beautiful and impressive structures the world has ever seen. The Great Pyramid of Khufu at Giza is one of the largest structures ever built and was the tallest building in the world until the 19th century. It is the only one of the Wonders of the Ancient World—a list made by travelers during ancient times of notable things to see around the world—that still stands. The pyramid stands 460 feet (140 m) high and its sides measure around 755 feet (230 m). It is made up of over two million limestone blocks that weigh an average of 2.5 tons each. In total, the ruins of 35 major pyramids can be found along the Nile River.

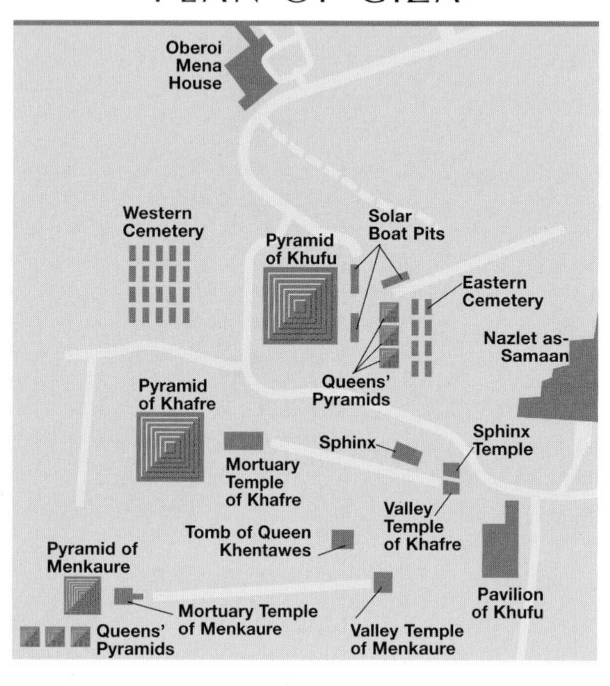


***Besides the three famous pyramids, Giza is also the site of temples and the tombs of the queens. Wooden boats that brought the pharaohs up the Nile were buried beside them to take them to the underworld.***

The Egyptian pharaohs built in stone because they wanted their tombs and temples to last for eternity. Architects copied features of older mud-brick temples. Stone columns and friezes were carved to resemble palm trunks and bunches of papyrus reeds. They were then decorated with natural motifs such as the lotus flower.

Temples such as the Great Temple of Amun at Karnak were designed as the setting for ritual processions in honor of the gods. An avenue lined with sphinxes (figures in the form of a reclining lion with the head and face of a king) led to a monumental pylon (gateway) and into an open courtyard surrounded by a colonnade. From here, the procession passed into a roofed chamber filled with a forest of massive columns, beyond which lay the dark and mysterious inner sanctuary.

The ancient Egyptians were accomplished sculptors. Tombs and temples were adorned with colossal statues of the pharaohs and huge stone sphinxes. These statues were traditionally supposed to guard the tomb or temples next to them. Temple gateways were traditionally flanked by statues of the king and a pair of obelisks—tall, tapering stone pillars covered with hieroglyphic inscriptions. Examples of the sculptors' art ranged from delicately executed life-size statues of the pharaoh and his queen to vast works such as the huge seated figures of Ramses II (1279–1212 B.C.) carved into the cliff face at Abu Simbel. Cats were popular subjects for small sculptures because the ancient Egyptians considered them sacred and valued them for their ability to protect grain from mice and rats. Sculptors used a variety of materials for their

***The French general Napoleon Bonaparte (1769–1821) is reputed to have sat in the shadow of the Great Pyramid and calculated that there was enough stone contained in the structure to build a 12-foot- (3.7-m-) high wall around France.***

*The legacy of ancient Egypt lives on to the present day in the form of some of the largest stone structures in the world. The Great Sphinx stands beside The Great Pyramid of Khufu in Giza.*

***The image on the left showing Ramses III and his son meeting the gods of the underworld, follows the strict painting rules of the Egyptian canon.***

## The Canon

The gods, pharaohs, queens, and court officials portrayed in Egyptian paintings can look very odd to modern eyes. Egyptian artists obeyed strict rules governing how artists depicted the human body. The ancient Egyptians noticed that, although people are many different sizes, the relative size of a limb to another was always the same—that, for example, the foot is three times the width of the hand. From this observation, they worked out a whole body of rules, or canon. In order to put this canon into practice, the Egyptian artist divided the painting into a grid of squares, each the width of the figure's foot. He then used this to work out the length of the figure's limbs, breadth of the shoulders, and so on.

This depiction of bodies can look very unnatural and twisted. The shoulders are shown from the front, while the face is shown in profile (from one side), and the figures always seem to have two left feet, with no little toes. Although they appear to be walking to left or right, in fact the Egyptians thought of the figures as moving out of the painting and toward the viewer.

The art of ancient Egypt changed little over the centuries. Artists depicted human bodies not as they saw them but as they knew them to be. Because they wanted to show as much of the human body as possible in one figure, they showed, for example, not only the profile but a full eye. In this way, they hoped to capture as much of the essence of the person they were portraying as possible and thus ensure his or her immortality.

small statues, including gold, bronze, ivory, and wood. The temples themselves were made of limestone and were often designed to resemble plants, with columns that were carved in the shape of reeds or palm trees.

## Egyptian Painting

Painting in ancient Egypt dates back to at least 3000 B.C. but reached its peak during the New Kingdom (c.1570–1070 B.C.). Tombs and palaces of this period contain colorful wall paintings of gods, pharaohs, plants, animals, scenes of everyday life, and images of the afterlife. Such paintings were not meant to be appreciated by the living but to ensure the immortality of the dead. It was believed that the scenes painted would come to life in the next world. The tomb owners therefore ensured that they were depicted looking their best and in attractive surroundings.

The Tomb of Nefertari, the first and favorite wife of Ramses II, lies in the Valley of the Queens near Luxor. It is decorated with some of the finest examples of pharaonic art in Egypt. A team of international conservationists has recently restored the paintings. Ancient Egyptian artists also turned their hands to the illustration of religious texts written on papyrus scrolls. Because papyrus is not damaged by rolling, many of these ancient texts still survive, thousands of years after they were written.

### Jewelry

The well-preserved tombs of Egypt's pharaohs have ensured that archaeologists have discovered many fine examples of ancient Egyptian jewelry. It was worn by both men and women. Gold was often used to make jewelry because it could be worked (bent into shape) easily and provided a rich color and texture. This precious metal was mined in Nubia to the south of Egypt. Semiprecious stones such as jasper or feldspar were often incorporated into the design. From the jewelry that has been discovered, it is clear that Egyptian jewelers had an advanced sense of design.

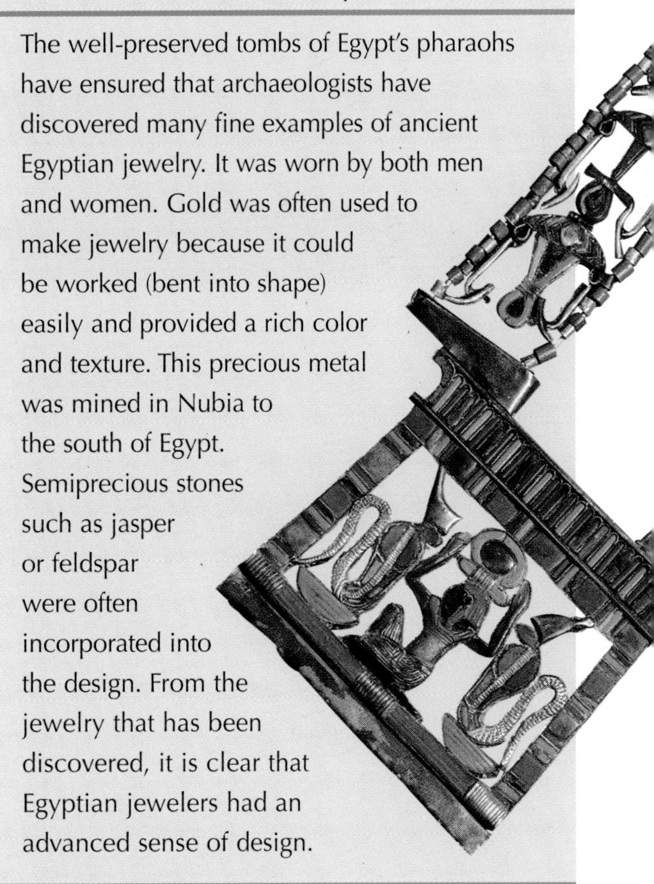

## Islamic Art and Architecture

*The sultans built a number of monuments in Cairo. Perhaps the most impressive is that of Sultan Qalawun. The madrasah, or religious college, pictured below, features windows with* mashrabiyyah *made of decorated inlaid stone.*

As one of the oldest Islamic cities in the world, Cairo contains some of the finest surviving examples of Islamic architecture. The Ibn-Tulun Mosque, built in A.D. 876 and 879, has outer courtyards enclosed by low, redbrick walls topped with decorative crenellations (battlements) and a large, pillared prayer hall with five aisles. Other important buildings include the mosque of Al-Azhar (970), the mosque of Al-Hakim (1010), and the mosque of Sultan Hassan (1356–1363).

Starting in the eighth century, Islamic teaching forbade the artistic representation of human and animal forms. Islamic teachers thought this would be an insult to God, who alone had the power to create living creatures.

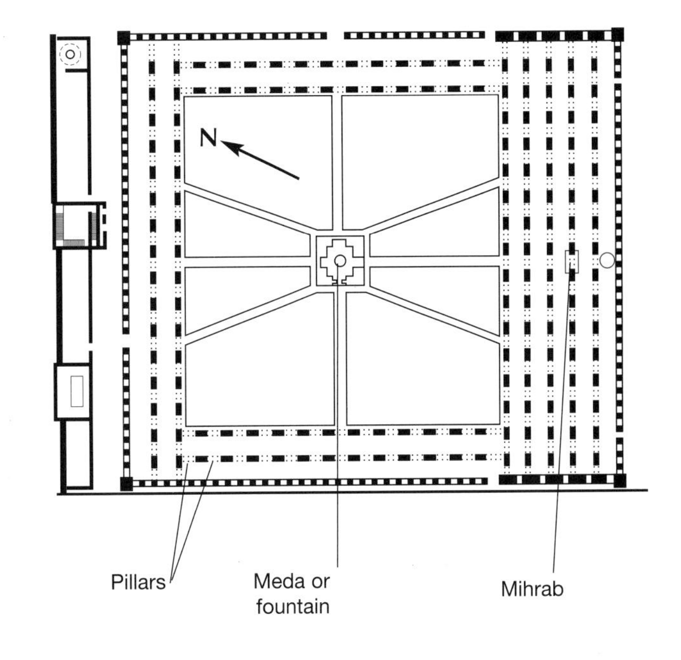


***The plan above shows the layout of the 9th-century Ibn Tulun Mosque at Cairo.***

So Islamic artists developed the use of intricate geometric designs. This new art reached its highest expression in the exquisitely carved wooden screens called *mashrabiyyah*. The *mashrabiyyah* covers the balcony window of the women's quarters in a traditional Muslim house or palace, so that the women can watch what is happening outside while remaining hidden from the outside world.

The art of calligraphy, or beautiful writing, was also used in Islamic decoration. The word Allah (God) and verses from the Koran in stylized Arabic script were interwoven in graceful, flowing patterns around the walls of mosques and palaces.

## The Arabic Script

ا ب ت ث ج ح خ د ذ ر ز

س ش ص ض ط ظ ع غ ف

ق ك ل م ن و ه لا ء ى ے

Islam's strict religious rules prevent artists from making images of living things. Muslims instead used calligraphy to decorate important texts such as the Koran. The Arabic script probably developed in around the fourth century A.D. from Nabatean, another Middle Eastern script. From the seventh century, the spread of Arabic was closely linked to the spread of Islam. The Arabic alphabet consists of 28 letters, all consonants, which are shown above. A system of smaller marks, developed in the eighth century, is sometimes added to indicate vowel sounds.

*In Egyptian mythology, writing was invented by Thoth, the god of wisdom and patron of scribes. Here he is shown as a baboon.*

## Writing and Literature

The ancient Egyptians wrote in hieroglyphics (from the Greek words meaning "sacred carving"). These were picture symbols that represented objects and people and also complex things such as ideas and sounds. Writing was used mostly for recording commercial transactions or to take down spells, prayers, and remedies. Generally only priests, officials, and scribes could read and write. Experts estimate that less than 1 percent of the population was literate.

Texts written in these symbols cover the walls and pillars of ancient Egyptian tombs, temples, and obelisks. Around A.D. 300, the Egyptians stopped using hieroglyphics and began to write using the Greek alphabet instead. After a time, no one was able to read hieroglyphics and their meaning was lost.

*Below is the seal of the great Egyptian queen Cleopatra VII, the last of the Ptolemaic Dynasty.*

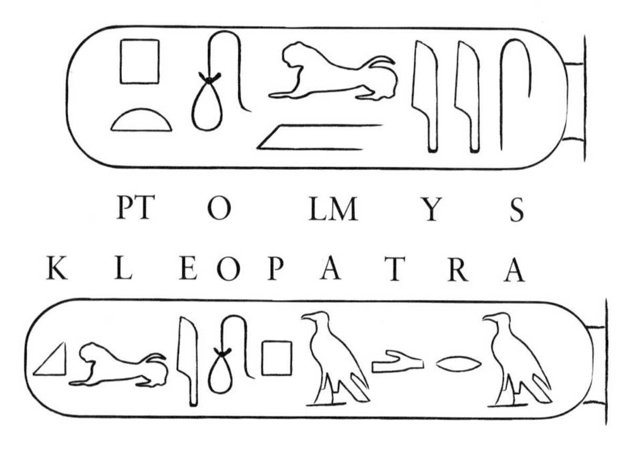


In 1799, the Rosetta Stone, an inscribed stone tablet, was discovered at Rosetta (Rashid), on the Mediterranean coast to the east of Alexandria. The stone was inscribed with a text written both in Egyptian hieroglyphics and Greek. This clue allowed a French scholar, Jean-François Champollion, to decipher the symbols. The discovery unlocked thousands of texts and massively increased knowledge and understanding of the ancient Egyptians.

Egypt has a long tradition of storytelling that goes back over a thousand years. Many of these narratives would be retold and embellished in different versions in different parts of the country. More recently, in the second half of the 20th century, Egypt's novelists have come under the influence of European writers. Egypt's most famous novelist is Naguib Mahfouz (born 1911). Many of his books have been translated into English and other languages, and he was awarded the Nobel Prize for Literature in 1988. His best-known work is a series of three novels—*Palace Walk*, *Palace of Desire*, and *Sugar Street*—known as *The Cairo Trilogy*. These novels, written in the late 1950s, record in colorful detail the lives of Egypt's urban poor. Mahfouz narrowly survived an attempt on his life by Muslim fundamentalists in 1994.

One of the most important and controversial women writers in Egypt is Nawal El-Saadawi (born 1931), best known for her novels *Point Zero*, *The Fall of the Imam*, and *Death of an Ex-Minister* and her nonfiction work *The Hidden Face of Eve: Women in the Arab World*. Her strong feminist views are disapproved of in conservative Egypt. She was imprisoned from 1981 to 1982, and some of her books are still banned in Egypt.

## Tomb Poems

During the 18th Dynasty, tomb paintings and reliefs often depicted lively banqueting scenes, with tumbling acrobats, musicians, and joyful dancers. They appear in rooms where relatives of the dead carried out rituals. The Egyptians' preoccupation with death did not mean that they did not enjoy earthly pleasures. On the contrary, awareness of death made them appreciate life even more. This poem was written in a tomb close to a banqueting scene:

*Follow your desire as long as you live,*
*Sprinkle your head with myrrh and put on fine linen,*
*Perfume yourself with the truly miraculous divine oils*
*Enjoy yourself as much as you can...*
*For a man cannot take his property with him*
*For of those who depart not one comes back again.*

## Music and Dance

Egyptians enjoy both classical Arabic music as well as modern sounds. Although still popular, classical music has considerably fewer listeners today than it did during the 1940s and 1950s. During these years, the singer Umm Kolthum most successfully captured the mood of the nation. Streets would become deserted as Egyptians everywhere gathered around the radio for her monthly live broadcasts. She was described as "the Mother of Egypt" or "Star of the Orient," and the devotion showed to her overshadowed the popularity of her male counterparts such as Abdel Halim Hafez. Although she died in 1975, her songs can still be heard throughout Egypt today.

***In Egypt, furniture was traditionally elaborately decorated. This detail from an ivory inlay dates to the Fatimid period (10th to 12th centuries A.D.).***

As in any culture, young people have developed their own type of music in Egypt. One of the pioneers of the modernization of Egyptian music was Ahmed Adawiyya. His work allowed *al jeel* ("the generation") and *shaabi* ("people") musical styles to evolve. *Al jeel* music combines traditional Egyptian rhythms with modern tracks of drums and vocals. *Shaabi* music has political lyrics that are often critical of modern Egyptian life; it is never played on Egyptian radio or TV.

## Movies

Egypt was one of the first Arabic countries to have its own movie industry, with the founding of Misr Studios in 1934. Today, Egypt is still the main producer of Arabic-language movies and TV shows. The country's most famous movie actor is Omar Sharif, who starred in many Hollywood movies in the 1960s and 1970s, including the classic *Lawrence of Arabia*.

## DAILY LIFE

Lifestyles in Egypt are dependent to a large extent on wealth and religion. A minority of Egyptians lead a very traditional way of life, in which women remain second-class citizens and wear concealing clothes. People live in very close-knit extended families, and the choice of marriage partners is heavily influenced by parents. At the other extreme are wealthy Egyptians for whom religion is of little importance compared to trips abroad and consumer goods. The lives of most Egyptians lie between these two extremes.

**The highest standard of living in Egypt is that of the Coptic Christians, who mainly live in the cities and are well educated.**

### Religious Life

Islam was founded by the prophet Mohammed, who was born in Mecca (now in Saudi Arabia) in about A.D. 570. The new religion arrived in Egypt in 642,

*Muslims must pray five times a day at specific times in the direction of Mecca. Here, a farmer stops work to pray.*

## The Islamic Calendar

The Arab world uses a different dating system—or calendar—from the West. The Muslim calendar is calculated according to the phases of the moon. It has 12 months of 29 or 30 days each. The beginning of each month coincides with the New Moon. As a result, the Islamic calendar year is 11 days shorter than the 365-day year of the Gregorian calendar used in the West. Islamic festivals fall 11 days earlier each year than when measured by the Gregorian calendar.

The years of the Muslim calendar are counted from A.D. 622, when the prophet Mohammed fled from Mecca to Medina, an event known in Arabic as the *Hegira.* New Year's Day 1421 A.H. (*anno Hegirae,* meaning "year of the *Hegira*") in the Muslim calendar is equal to April 6, 2000, in the Gregorian calendar.

only ten years after Mohammed's death, and was widely embraced, in part because of the persecution of Egyptian (Coptic) Christianity by the Byzantine empire. Egypt also adopted the Arabic language, in which the Koran (al-Qur'an), the holy book of Islam, was written. During the Fatimid dynasty (969–1171), Egypt became the center of the Islamic world, and the founding of the al-Azhar mosque and university made Cairo an important center of Islamic scholarship.

*Muslims are called to prayer by a* **muezzin,** *who cries from a tower on a mosque known as a* **minaret.**

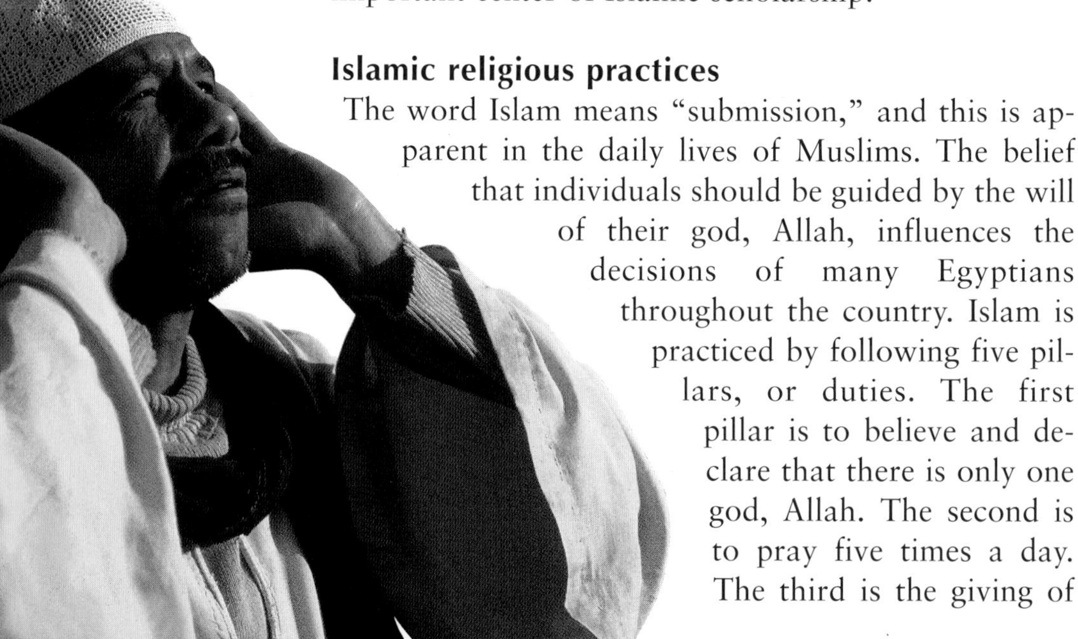

### Islamic religious practices

The word Islam means "submission," and this is apparent in the daily lives of Muslims. The belief that individuals should be guided by the will of their god, Allah, influences the decisions of many Egyptians throughout the country. Islam is practiced by following five pillars, or duties. The first pillar is to believe and declare that there is only one god, Allah. The second is to pray five times a day. The third is the giving of alms, or donations, to the poor and to help develop the Islamic faith. The fourth pillar is the yearly fast during the daylight hours of the Islamic month of Ramadan. Finally, Muslims must try to make a pilgrimage to Mecca, if they can afford it, at least once during their lifetime. This pilgrimage is known as the *hajj*. Observing these rituals plays a large part in the lives of devout Muslims and strongly influences the way they live. Muslims must pray at set times of the day, beginning at sunrise, and the fast they undertake each year during Ramadan can make Muslims feel weak and tired in that month.

### The *Hajj*

The *hajj* is the Muslim pilgrimage to Mecca, the birthplace of Mohammed and the holiest site in Islam. Once a pilgrim reaches the outskirts of Mecca, he must change into pilgrim's clothing—simple garments worn by rich and poor alike. Mecca's al-Haram Mosque holds the Ka'bah, which according to tradition, was built by Abraham and his son Ishmael as a replica of God's house in heaven.

## The Coptic Church

While the official religion of Egypt is Islam, Coptic Christians form a significant religious minority of about 6 percent in the country. The head of the Coptic church is known as the Patriarch of Alexandria and All Egypt. His residence is in Cairo. Egyptians adopted Christianity soon after Saint Mark began preaching the gospel there in around A.D. 40. Alexandria was a Christian city by the end of the first century A.D., and Christianity became the official religion of Egypt in the fourth century.

The Coptic church has different beliefs from Eastern Orthodox churches and from the Roman Catholic church; these differences emerged in the fourth century A.D. Conflict with the Orthodox church in Constantinople during the fifth century led to the isolation of the Egyptian church. After the Arab conquest of Egypt in the seventh century, this isolation increased. In the centuries following the invasion, those Egyptians who did not adopt the Muslim religion held on to their Christian beliefs. Today the Copts form a distinct cultural group from the Muslim majority.

**HOW EGYPTIANS SPEND THEIR MONEY**

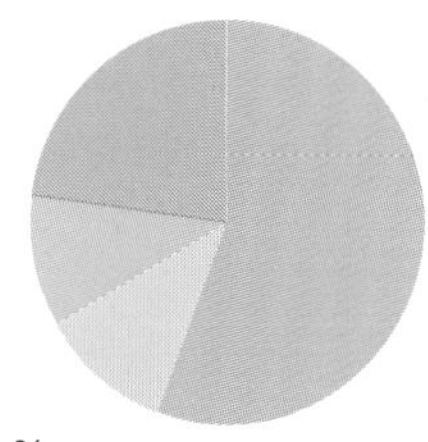

%

- 55.7 Food and drink
- 10.9 Clothing
- 10.5 Housing
- 22.9 Other

***The relative poverty of many Egyptians means that food and shelter are still the main expenses.***

Since the early 1980s there have been increasing tensions between Muslim fundamentalists and Coptic Christians. A church in Cairo was bombed in 1981, and two Coptic priests were murdered at the monastery of Deir al-Muharraq in 1994. Since 1992, an estimated 500 people have died in the violence.

## Food and Drink

The traditional cuisine of Egypt is a mixture of North African, Mediterranean, and Middle Eastern influences. Popular foods include rice, beans, chickpeas, and eggplant. Most meat dishes are based on lamb and chicken, and fish is also popular. Roasted pumpkin or watermelon seeds are eaten as a snack.

The most popular dish in Egypt is probably *fuul*—cooked fava beans (brown beans) dressed with olive oil, lemon juice, and garlic. It is eaten at breakfast, lunch, and dinner, or in a sandwich as a snack. The most common meat dishes are kebab (chunks of lamb speared on a skewer and grilled over charcoal) and *kofte* (meatballs

***Egyptian males enjoy spending time at the coffeehouse (*ahwa*). Here they meet with friends, read the newspapers, and play dominoes or backgammon.***

## Fuul

This is one of the most popular Egyptian dishes, eaten at breakfast or lunch or as a side dish to a main meal. Serve with pita bread and a tomato and onion salad.

**Serves six people**

1½ lb (700g) dried fuul or broad beans
6 garlic cloves, crushed
4 tbs finely chopped parsley
2 lemons, quartered
Olive oil
Salt and black pepper
Cayenne pepper

Put the beans in a large saucepan, cover with water, and leave to soak overnight. Next day drain the beans, cover them with fresh water and bring to a boil. Reduce heat and simmer for 2 hours until the beans are soft but not mushy. Remove from the heat and add the crushed garlic. Spoon the mixture into a serving bowl and sprinkle the parsley on top, garnish with the lemon quarters and season with the oil, salt, and pepper.

of minced lamb seasoned with onion, parsley, and cumin, also grilled over charcoal). A popular street snack is *ta'amiya*, which are deep-fried patties made of chickpeas seasoned with garlic and cumin served with tahini (pulped sesame seeds) and pita bread.

Egyptians have a sweet tooth and enjoy a wide range of sticky desserts and pastries similar to those enjoyed in Turkey. These include *baqlawah* (alternating layers of filo pastry and chopped almond and pistachio nuts saturated with sugar syrup), *basboosa* (semolina cake, baked with butter and drenched with syrup), and *kounafa* (nests of pastry thread filled with chopped nuts and soaked in honey).

Tea (*shay*) and coffee (*ahwa*) are popular beverages. Tea is brewed very strong and served in small tulip-shaped glasses with sugar but no milk. Coffee is also made very strong and sweet and served in small porcelain cups. Freshly squeezed juice—orange, lime, pomegranate, sugarcane, strawberry, or tamarind—is sold at street stands in most Egyptian towns.

## National Holidays and Festivals

Muslims all over the world follow their own special calendar for religious events (*see* box, p.106). The Islamic calendar is divided into months based on the phases of the moon. This means that the months on the Islamic calendar are irregular and that holidays have to be calculated each year. The beginning of each month is decided when the new moon is seen by specially chosen Muslim scholars. The calendar is 11 days shorter than the Gregorian calendar used in the West and dates therefore move backward 11 days every year. For example, the month of Ramadan, which starts on November 16, 2001 in the Gregorian calendar, will begin on November 5 in the Islamic calendar.

The following public holidays are on fixed dates each year:

| | |
|---|---|
| January 1 | New Years' Day |
| January 7 | Christmas |
| April 25 | Anniversary of Liberation of Sinai |
| May 1 | May Day |
| July 23 | Anniversary of the the Revolution |
| October 6 | National Day |

The *Ras as-Sana* and *Moulid an-Nabi* religious festivals (*see* National Holidays and Festivals) are also public holidays, but fall on a different date each year. The Coptic Church also celebrates Easter in March or April.

**National Holidays and Festivals**

The most important holidays in the Egyptian year are Islamic and Coptic religious festivals. Islamic festivals are celebrated according to the Muslim calendar (*see* above). The most important event is the ninth month of the Islamic calendar, called Ramadan. Muslims believe that this is when the Koran was revealed to the prophet Mohammed by Allah (the Muslim name for god). For the duration of Ramadan, Muslims observe a fast during the hours of daylight and do not eat, drink, or smoke from dawn to dusk. Fasting during Ramadan demonstrates a Muslim's willingness to surrender to the will of Allah—this is the most important belief in the religion of Islam.

The end of Ramadan is marked by the festival of *Eid al-Fitr*, a three-day public holiday of feasting and celebration. Other important public holidays are *Ras as-*

## How to Say...

While Standard Classical Arabic is pretty similar throughout the Arab-speaking countries, the language used everyday in Egypt, Egyptian Colloquial Arabic (ECA), differs so significantly that it is practically another language. Generally, the more informal the topic of conversation, the more informal the Arabic. ECA is notoriously difficult to learn. Nevertheless, here are a few basic phrases to try. An apostrophe indicates a glottal stop, a sound produced by closing the throat.

**Basic Expressions**

Do you speak English? *Enta bitikallim ingilīzī?* (to a male) / *Enfi bitikallimī ingilīzī?* (to a female)
Yes *Aywa, Na'am* (more formal)
No *La'*
I understand *Ana fāhem*
I don't understand *Ana mish fāhem*

Hello (greeting) *Salām alēkum*
Hello (response) *Wa 'alēkum es salām*
Good morning (greeting) *Sabāh al-khēr*
Good morning (response) *Sabāh an-nūr*
Good evening (greeting) *Misa' al-khēr*
Good evening (response) *Misa' an-nūr*
Goodbye *Ma'as salāma*

Please *Min fadlak* (to a male) / *Min fadlik* (to female)
Thank you *Shukran*
You're welcome *'Afwan, al-'affu*
No thank you *La' shukran*
Sorry *'Assif*

How are you? *Izzayak?* (to male) / *Izzayyik* (to female)
Fine, thank you/good *Kwayyis, il-Hamdu lillah* (to male) / *Kwaysa, il-Hamdu lillah* (to female)
What's your name? *Ismak ēh?* (to male) / *Ismīk ēh* (to female)
My name is ... *Ismī ...*
Where are you from? *Min ayna tati?*
I'm from the United States *Ayna min al-wilayat al-mottahidah*
I would like ... please *Orid ... min fadlīk*
How much does this cost? *Bikam haza?*
Do you have...? *Fi 'andak...?*

**Numbers**

One *Wahid*
Two *Itnein*
Three *Talata*
Four *Arba'a*
Five *Khamsa*
Six *Sitta*
Seven *Sab'a*
Eight *Tamanya*
Nine *Tis'a*
Ten *'Ashara*

***Ownership of consumer goods is still low in Egypt when compared to the West.***

**WHAT DO EGYPTIANS OWN?**

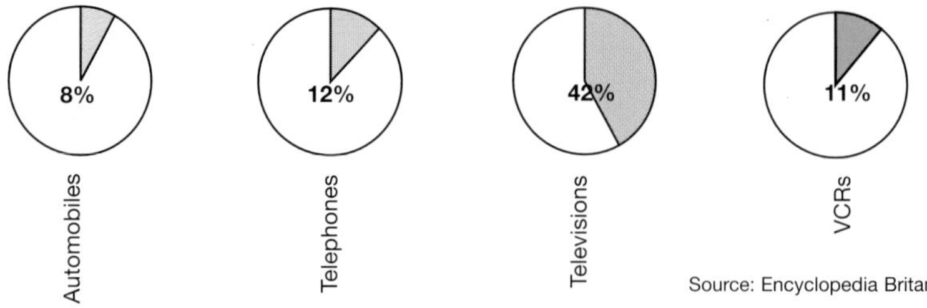

| Automobiles | Telephones | Televisions | VCRs |
|---|---|---|---|
| 8% | 12% | 42% | 11% |

Source: Encyclopedia Britannica

*Sana* (the Muslim New Year) and *Moulid an-Nabi* (the birthday of the prophet Mohammed). Several other festivals (*moulid*) celebrating the birthdays of local saints or holy men are held throughout the year and generally last for up to a week.

Coptic Christian festivals, like those of the Eastern Orthodox churches, follow the Julian calendar, which is 14 days behind the Gregorian calendar used in the West. So in terms of the dates celebrated in the West, Christmas Day is celebrated on January 7, Epiphany on January 19, and the Annunciation on March 23. Easter is the most important festival of the Coptic year.

**EDUCATIONAL ATTENDANCE**

|  |  |
|---|---|
| Further (university) | 20% |
| Secondary (high school) | 81% |
| Primary | 97% |

Source: Government of Egypt

***Government claims of school attendance figures are much higher than in reality.***

## Sports and Leisure

The most popular spectator sport in Egypt is soccer. Matches played by the major Cairo teams Al-Ahly and Zamalek draw huge crowds at the stadiums and large audiences on television. The Egyptian national soccer team is one of the best in the Middle East, and qualified for the World Cup in 1998. Egyptian men spend much of their leisure time in tea and coffeehouses, chatting, smoking, and playing chess or backgammon.

## Education

Primary and secondary school education in Egypt is free, and attendance is compulsory for children aged six to 12. Children go to primary school for six years. This is followed by two more stages: preparatory level for three years and secondary level for a further three years. Many children do not continue their education

## Women in Egypt

In 1899, Qasim Amin, an Egyptian writer, published a work blaming Egyptian society's backwardness on its oppression of women. Since then, the situation for women in Egypt has improved, but not dramatically. Egypt has led the way in women's rights among the Arab nations. In the 1920s, the veiling of women became unfashionable. In 1956, women won the right to vote and female representatives entered parliament.

Recently, however, there has been a swing back to traditional Islamic values, and women are once again dressing conservatively. Women are still treated differently from men under Islamic law. If a man wishes to divorce his wife, he simply has to utter an oath. For women, divorce involves a lengthy legal process.

beyond primary level, especially in rural areas where they are needed to help their families work on the land. As a result, around 50 percent of Egyptians over the age of 15 cannot read or write. Two-thirds of these are women. The government has recently taken steps to encourage preparatory and secondary education in order to provide a trained workforce to work in industry.

Besides the state education system, there are schools associated with the al-Azhar University in Cairo. These provide primary and secondary education with special emphasis on the study of the Koran and on traditional Islamic subjects. The Egyptian university system is respected across the Arab world.

**Health care in Egypt is basic, with one hospital bed for every 500 people and one doctor per 556 people. This compares with one doctor per 400 people in the United States.**

## Health and Welfare

The government has created a network of health centers in rural areas to help combat common diseases. Children receive compulsory vaccinations against diphtheria, tuberculosis, and polio. The tropical disease schistosomiasis, also known as bilharziasis, is a serious health problem throughout Egypt, hindering development in young children and reducing their life expectancy. The disease is caused by a tiny parasitic flatworm and is transmitted through contact with contaminated freshwater, by drinking, swimming, or washing. Malaria is still a serious problem in the Nile Valley. Islamic medical centers are now replacing much of the state system at the local level.

## Media

Cairo is the most important publishing and media center in the Arabic world. Criticism of the government by Islamic fundamentalists led to new controls over the press in 1995. These were reinforced in 1998 after the opposition criticized the security clampdown. The press is subject to a degree of government control and censorship. The primary Arabic-language daily newspaper is *Al-Ahram*, which also publishes weekly news summaries in English and French. *The Middle East Times* is less academic than *Al-Ahram*. *The Egyptian Gazette* is an English-language daily published in Cairo.

There are three national television channels run by the government-owned Egyptian Radio and TV Corporation. Channel 2 screens a variety of programs and movies in English. More recently, local governorates have set up their own stations. Nile TV, based in Cairo, carries news stories and current-affairs features in English and French. Various international channels are available via satellite TV, and cable TV is becoming available. Due to the relatively low level of TV ownership, radio also attracts many listeners.

**Egypt's daily newspaper circulation is low at 38 per 1,000 people, compared with 212 per 1,000 people in the United States.**

## The Bedouin

The bedouin (*badawin*, meaning "desert-dweller") are Arab tribespeople who move from place to place in the deserts of Arabia, Jordan, Syria, Israel, and parts of North Africa. In Egypt, bedouin tribes are found in both the Eastern and Western deserts, but the largest numbers live in Sinai. There are about 500,000 bedouin living within Egypt's borders.

The traditional bedouin lifestyle is nomadic, and a man's wealth is measured in camels and children. Small bands of people, with their herds of camels and goats, trek through the desert from waterhole to waterhole in search of scant pasture for their livestock. When the grass at one spot is gone, they take down their black goat's-wool tents and move on, living on a sparse diet of grain, dates, goat's milk, and occasional feasts of mutton. They also shoot and eat game such as quail and antelope.

This wandering lifestyle is under threat, however. The bedouin are coming under increasing pressure from the Egyptian government to settle in one place. In addition, traditional bedouin areas, such as the oases of the Western Desert, are being developed as land for the excess population of the Nile Valley. Four-wheel drive vehicles are replacing camels as the "ship of the desert," and many bedouin are forsaking their tents for permanent settlements and losing traditional skills.

# The Future

*"Globalization would lead to an improvement in the standard of living in developing countries."*

Hosni Mubarak, current Egyptian president

As Egypt enters its seventh millennium of recorded history, it faces a number of challenges, including rapid population growth, overcrowding, water shortages, and the rise of Islamic fundamentalism.

## Population Explosion

Egypt is the second most populous nation in Africa after Nigeria. The population was estimated at 66 million in 1998 and is growing at an annual rate of around 2.1 percent—an increase of almost 1.4 million every year. The main population problem, however, is the population density. Because almost everyone lives in the cultivated areas of the Nile Valley and Delta, which comprise only 6 percent of the country's land area, one of the highest population densities in the world has been created there—4,000 persons per square mile (1,558 per sq km).

This explosive population growth—the total is projected to reach almost 77.5 million by 2010—is putting an enormous strain on the country's resources. Egypt has an overwhelmingly young population. Almost 40 percent of Egyptians are under 15 years old. Despite improvements in health care, many children die in infancy. The infant mortality rate was 67.5 per 1,000 live births in 1997. Schools and health services are trying to cope with ever-increasing numbers of children, and there is a serious overcrowding problem in Egyptian

***The average household in Egypt has 4.3 people. Many couples continue to live with the husband's parents even after they are married.***

FACT FILE

- During the period 1996–1997 Egypt was second only to Israel in terms of the amount of financial aid it received from the United States.
- Average life expectancy in Egypt currently stands at 62 years—60.39 for men and 64.49 for women—higher than the average for Africa. Infant mortality is high, at 67.5 deaths for every 1,000 births.
- Egypt has the largest diplomatic service in the Arab world, and one of the most widely respected. The former U.N. Secretary-General Boutros Boutros–Ghali is an Egyptian.

cities. Cairo is blighted by slums and squatter settlements, and hundreds of thousands of people live in makeshift homes in the medieval cemeteries in the eastern suburbs—the so-called Cities of the Dead.

The Egyptian government is trying to deal with this problem by encouraging family planning, and by increasing the habitable area of the country. But both the increasing population and the expansion of cultivated land is putting more pressure on Egypt's most precious natural resource—water.

## The Fundamentalist Challenge

**Egypt's relations with Iran are tense because Iran actively supports the fundamentalist Islamic groups that oppose the Egyptian government. Iran accuses Egypt of being a corrupt state, heavily under U.S. influence.**

Since the 1970s the rise of Islamic fundamentalism has led to increased social tensions, especially between Muslims and Coptic Christians. This led to terrorist attacks and outbursts of violence. In the 1990s, these groups targeted Egypt's tourist industry in an attempt to influence the government, attacking tour buses and Nile cruise ships with guns and bombs. In 1996, 18 tourists were killed when gunmen opened fire on a Cairo hotel, and nine tourists and their Egyptian driver died when their bus was attacked on September 18, 1997. In the worst incident, on November 17, 1997, 58 tourists were murdered by *el-Jihad* terrorists at the Temple of Queen Hatshepsut in Luxor.

These attacks have caused a serious downturn in the Egyptian tourist industry, which is an important sector of the economy. The government has responded by cracking down on fundamentalist groups, and increasing security in tourist areas.

## Precious Water

For a desert nation, water is as precious as gold or oil. Egypt relies for its water almost entirely on the Nile, which provides around 97 percent of its water resources (the rest comes from groundwater and recycling). In 1998, Egypt's water consumption almost matched its available resources. By 2025 the increasing

## Egypt's Water Resources and Consumption

| Resources | (billions of cu m per year) | Consumption | (billions of cu m per year) |
|---|---|---|---|
| Nile water | 55.5 | Agriculture | 53.1 |
| Recycled agricultural drainage water | 4.5 | Drinking water | 4.5 |
| Treated sanitary drainage water | 0.7 | Industry | 7.5 |
| Groundwater in Nile Valley and Delta | 4.8 | | |
| Deep subterranean water | 0.6 | | |
| Totals | 66.1 | | 65.1 |

population will reduce the amount of water available from around 35,300 cubic feet (1,000 cu m) per head of population to 22,770 cubic feet (645 cu m).

As demands on water resources increase, friction between Ethiopia, Sudan, and Egypt is likely. All these countries have a claim on the Nile. In addition, there may be conflict between Egypt and Libya over the use of water that lies beneath the Western Desert.

Egypt therefore faces conflicting demands. On one hand there is the desire for more agricultural land—which will need water for irrigation—to feed and house a growing population. On the other there is the need to conserve and control the limited water resources available. The government is adopting three measures: first, making more efficient use of available water; second, conserving current water resources; and third, developing new water resources, such as deep aquifers (a layer of porous rock that holds water), and desalination plants to purify saltwater that is available on the country's coasts.

# Almanac

## POLITICAL

**Country name:**
Official long form*:* Arab Republic of Egypt
Short form: Egypt
Local long form: *Jumhuriyat Misr al-Arabiyah*
Local short form: *Misr*

**Nationality:**
noun: Egyptian (s)
adjective: Egyptian

**Official language:** Arabic

**Capital city:** Cairo

**Type of government:** republic

**Suffrage (voting rights):** Compulsory for everyone 18 years and over

**National anthem:** "Biladi!" ("My Homeland")

**National holiday:** July 23

**Flag:**

## GEOGRAPHICAL

**Location:** Northern Africa; latitudes 27° north and longitudes 30° east

**Climate:** desert, hot, dry summers with moderate winters

**Total area:** 386,900 square miles (1,002,071 sq. km)
land: 99%
water: 1%

**Coastline:** 1,470 miles (2,450 km)

**Terrain:** vast desert plateau interrupted by Nile Valley and Delta

**Highest point:** Mount Catherine (Jebel Katrina), 8,688 feet (2,642 m)
**Lowest point:** Qattara Depression, -435 feet (-133 m)

**Natural resources:** petroleum, natural gas, iron ore, phosphates, manganese, limestone, gypsum, talc, asbestos, lead, zinc

**Land use** (1993 est.):
arable land: 2%
other: 98%

**Natural hazards:** periodic droughts, earthquakes, flash floods, landslides, sandstorms

## POPULATION

**Population:** 66 million

**Population density:** 171 people per square mile (66 per sq. km)

**Population growth rate** (1999 est.): 2.1%

**Birthrate** (1999 est.): 26.8 births per 1,000 of the population

**Death rate** (1999 est.): 8.27 deaths per 1,000 of the population

**Sex ratio** (1999 est.): 103.1 males per 100 females

**Total fertility rate** (1999 est.): 3.3 per woman in the population

**Infant mortality rate** (1999 est.): 67.5 deaths per 1,000 live births

**Life expectancy at birth** (1999 est.):
total population: 62.39 years
male: 60.39 years
female: 64.49 years

**Literacy:**
total population: 51.4%
male: 63.6%
female: 38.8%

## ECONOMY

**Currency:** Egyptian pound (guinay) (£E); 1 £E= 100 piastres

**Exchange rate** (1999): $1 = £E3.39

**Gross national product** (1999): $72.1 billion (42nd-largest economy in the world)

**Average annual growth rate** (1990–1997): 4%

**GNP per capita** (1999 est.): $1,200

**Average annual inflation rate** (1990–1997): 11.5%

**Unemployment rate** (2000): 10%

**Exports** (1997): $4.9 billion

**Imports** (1997): $14.7 billion

**Foreign aid received** (1996): $2.4 billion

**Human Development Index**
(an index scaled from 0 to 100 combining statistics indicating adult literacy, years of schooling, life expectancy, and income levels):
61.2 (U.S. 94.2)

# TIME LINE—EGYPT

**World History** | **Egyptian History**

## *c.* 50,000 B.C.

***c.* 40,000** Modern humans—*Homo sapiens sapiens*—emerge.

## *c.* 5000 B.C.

***c.* 3500 B.C.** Early settlement in Mesopotamia (modern Iraq).

**2325** B.C. Sargon builds an empire in Babylonia and Assyria.

***c.* 5000 B.C.** First agricultural settlements in Egypt.

**3100 B.C.** Upper and Lower Egypt united.

**2686–2184 B.C.** Period of Old Kingdom.

**2040–1782 B.C.** Period of Middle Kingdom.

## *c.* 1500 B.C.

**1200–1000 B.C.** Phoenicians rise to power in the Mediterranean.

**753 B.C.** City of Rome founded.

**559 B.C.** Rise of Persia.

**1570–1070 B.C.** Period of New Kingdom.

**674–600 B.C.** Assyrian and Babylonian attacks repulsed.

**440 B.C.** Greek historian Herodotus visits Egypt.

## *c.* 400 B.C.

**c. A.D. 30** Death of Christ in Palestine.

**146 B.C.** Greece comes under Roman rule.

**A.D. 30** Egypt becomes a Roman province.

**52–30 B.C.** Reign of Cleopatra VII.

**332 B.C.** Alexander the Great seizes Egypt.

## A.D. 100

**1526** Foundation of Mughal empire.

**1288** Ottoman state founded in Turkey.

**1065** Start of the Crusades.

**429–535** Era of Vandal kingdom in North Africa.

**1516** Start of Ottoman rule of Egypt, Syria, and Arabia.

**1171** Saladin conquers Egypt.

**1168** Crusaders threaten Egypt.

**641** Arabs conquer Egypt.

**200–300** Rise of Coptic Christianity.

## 1700

**1867–73** David Livingstone searches for the source of the Nile.

**1792–1815** Napoleonic Wars in Europe.

**1775–83** American War of Independence.

**1860** Thomas Cook begins the first organized tours of Egypt.

**1811** Mohammed Ali takes control.

**1798** Napoleon attacks Egypt.

**1850**

**1880s onward** European powers (particularly France and Britain) conquer most of Africa.

**1914–1918** World War I.

**1917** Revolution in Russia leads to the rise of communism.

**1869** Suez Canal opened.

**1882** British begin occupation of Egypt.

**1914** Egypt becomes British protectorate.

**1919** Nationalist revolt begins.

**1920**

**1922** Last Ottoman emperor deposed.

**1929** Wall Street crash leads to world economic recession.

**1939–1945** World War II.

**1945–65** Collapse of European colonialism and the rise of independence movements in former colonies.

**1922** Egypt becomes independent.

**1924** Wafd (independence) party wins landslide in general elections.

**1936** King Farouk comes to the throne.

**1942** Support for Germans during World War II in order to end British rule.

**1948–1949** Egypt and its Arab neighbors go to war with newly formed State of Israel.

**2000** The West celebrates the millennium.

**1990** Gulf War.

**1989–91** Fall of communism in Eastern Europe.

**Late 1970s** Rise of Islamic fundamentalism in North Africa and Middle East.

**1973** Massive hike in oil prices by Arab states leads to recession in the West.

**1997** Terrorists kill tourists at Luxor.

**1990** Arab League returns to Cairo.

**1981** Anwar Sadat assassinated. Hosni Mubarak becomes president.

**1979** Egyptian-Israeli peace deal signed at Camp David. Return of Sinai to Egypt.

**1973** Yom Kippur War with Israel.

**1971** Opening of Aswan High Dam.

**1970** Anwar Sadat becomes president.

**1970**

**1969** First man lands on the moon.

**1956** Hungarian uprising suppressed by Soviet troops.

**1950–53** Korean War.

**1967** Six-Day War with Israel.

**1956** Suez Crisis.

**1954–70** Presidency of Abdel Nasser.

**1952** Military coup deposes the king.

**1950**

# Glossary

**Allah:** The name used by Muslims to refer to God.
**Arab:** A people originating in the Arabian peninsula who now live throughout northern Africa and southwestern Asia.
**Aswan High Dam:** A vast dam in southern Egypt, built in the 1960s and completed in 1971, which now provides Egypt with huge hydroelectric resources and helps regulate the flooding of the Nile River.
**Ayyubid:** Important dynasty that ruled Egypt in the 12th and 13th centuries.

**bedouin:** A traveling people who live in the deserts of North Africa and the Middle East.

**Coptic:** A branch of Christianity originating in Egypt that broke off from the Orthodox church in the third century A.D.
**capitalism:** An economic system based on supply and demand and private ownership of businesses and industry.
**catacombs:** Underground burial chambers.
**Christianity:** Religion based on the teachings of Jesus Christ.
**colonialism:** Control of one country or people by another in a subordinate role.
**communism:** A social and political system based on a planned economy in which goods and land are owned by everyone and in which there is no private property.
**constitution:** A written collection of a country's laws, its citizens' rights, and principal beliefs.

**delta:** An area at the mouth of a river, where the river splits into many different streams, creating a marshy landscape.
**democratic:** Description of a process or state where the people choose their government by free elections, and where supreme power is held by the people.

**export:** A product that is sold to another country.

**Fatimid:** Dynasty or family that ruled Egypt from 969 to 1171, believing themselves to be descended from Fatima, the daughter of Mohammed.
**fundamentalism:** Religious belief founded on strict adherence to a set of basic principles.

**hieroglyphs:** A form of writing in small pictures or symbols used by the ancient Egyptians.
**hydroelectric power:** Energy produced by the motion of water.

**irrigation:** The watering of plants by artificial channels, pipelines, or other means.
**industrial:** Describing an economy based on developed industries and infrastructure rather than on agriculture.
**Islam:** Religion based on the teachings of Mohammed.

**Koran:** The holy book of Islam.

**minaret:** A tall tower attached to a mosque from which the Muslim faithful are called to prayer.
**Mohammed:** Seventh-century founder of the religion of Islam.
**mosque:** The house of worship in the religion of Islam.
**mummy:** A dead body wrapped in cloth bandages and then placed in a sarcophagus (coffin).
**Muslim:** A follower of Islam.

**nationalization:** The placing of private industry under public control.

**oasis:** An area of fertile green in the middle of a desert.
**obelisk:** A tall, pointed, tapering stone column.
**open market:** A market without tariffs (a charge imposed by government to limit certain types of trade) or other trade barriers.
**Ottoman:** Turkish empire, founded in the 14th century, ruled Egypt for over 300 years.

**Palestine:** The historical name for the country that previously occupied the territory of modern-day Israel.

**papyrus:** A type of tall grass used in ancient Egypt (and other parts of the ancient world) to make a form of early paper.
**pharaoh:** Ruler or king of ancient Egypt.

**Ramadan:** Ninth month of the Islamic calendar; held sacred by Muslims as the time in which the Koran was revealed by Allah to Mohammed.
**republic:** A government in which the citizens of a country hold supreme power and where all citizens are equal under the law.

**sarcophagus:** A stone coffin.
**Sinai:** Peninsula in the east of Egypt, largely covered by desert and arid mountains.
**socialism:** A system of government where goods and industry are publicly owned and where the economy is planned.
**Suez:** Important Egyptian shipping canal, linking the Indian Ocean with the Mediterranean Sea. Also the name of the port and administrative region neighboring the canal.

**trade surplus:** The situation in a country's economy when the value of the goods sold overseas (its exports) is greater than goods bought from overseas (its imports).

**underworld:** A place where the Egyptians believed that they went after death.

# Bibliography


**Major Sources Used for This Book**
Clayton, Peter A. *Chronicle of the Pharaohs.* London: Thames and Hudson, 1994.
Daly, M.W (ed.). *The Cambridge History of Egypt.* Cambridge University Press, 1999.
Empreur, Jean-Yves. *Alexandria Rediscovered.* New York: George Braziller, 1998.
*CIA World Factbook 1998* (www.odci.gov/cia/publications/factbook)

**General Further Reading**
*Encyclopedia of World Cultures* ed. Lynda Bennett. Boston: G.K. Hall & Co., 1992.
*World Reference Atlas.* London: Dorling Kindersley, 2000.
*The Kingfisher History Encyclopedia.* New York: Kingfisher, 1999.
*Student Atlas.* New York: Dorling Kindersley, 1998.
*The World Book Encyclopedia.* Chicago: Scott Fetzer Company, 1999.

**Further Reading About Egypt**
Brown, Deni and Steedman, Scott. *Ancient Egypt.* New York: Dorling Kindersley, 1996.
Compoint, Stephane and Riche, William, L.A. *Alexandria: The Sunken City.* London: Weidenfeld and Nicholson, 1997.
Morrison, Ian A. *Egypt (World in View).* Austin, TX: Steck-Vaughn, 1991.
Murname, William J. *The Penguin Guide to Ancient Egypt.* New York: Penguin, 1997.
*What Life Was Like on the Banks of the Nile.* Alexandria, VA: Time-Life Books, 1997.

**Some Websites About Egypt**
www.sis.gov.eg (Egyptian State Information Service)
www.touregypt.net (Egyptian Tourist Board)

# Index

**Page numbers in *italics* refer to pictures or their captions.**

# Acknowledgments

**Cover Photo Credits**
**Corbis:** Roger Wood (man sipping drink); **Corbis:** Neil Beer (Tutankhamun death mask); **Corbis:** Jonathan Blair (Nile River)

**Photo Credits**
**AKG London:** 60, 66
**Corbis:** 71, Archivo Iconografico, S.A. 48, 49; Yann Arthus-Bertrand 19, 43; Bettmann 41, 72; Jonathan Blair 6; Bojan Brecelj 30; Dean Conger 90; Gianni Dagli Orti 44, 55, 56, 69, 98; Hans Georg Roth 38; Robert Holmes 88; Peter Johnson 31; Steve Kaufman 24; Ludovic Maisant 93; Joe MacDonald 29; Ali Meyer 68; Richard T. Nowitz 17, 116; Christine Osborne 33, 106; Fulvio Roiter 113; Jeffrey L. Rotman 115; Kevin Schafer 28; The Purcell Team 80; Vanni Archive 100; Nik Wheeler 16; Staffan Widstrand 105; Adam Woolfitt 1, 97
**Egyptian State Tourist Office, London:** 21, 22, 37, 47, 50, 62; **Hutchison Library:** Carlos Freire 40; Reditt 20; Liba Taylor 108; **Robert Hunt Library:** 73, 75, 76; **TRIP Photo Library:** 83, 84; T. Bongar 18, A. Ghazzal 12, 91; J. Pilkington 52; H. Rogers 86;
**Werner Forman Archive:** British Museum, London 54; Egyptian Museum. Berlin 102; Egyptian Museum, Cairo 57, 99; Museum für Islamische Kunst, Berlin 104; E. Strouhal 59